Mission
Jupiter

Daniel Fischer

Mission Jupiter

The Spectacular Journey
of the *Galileo* Spacecraft

C

COPERNICUS BOOKS
An Imprint of Springer-Verlag

Originally published as *Mission Jupiter: Die spektakuläre Reise der Raumsonde Galileo,*
© 1998 by Birkhäuser Verlag, Basel, Switzerland.

© 2001 Springer-Verlag New York, Inc.

Published in the United States by Copernicus Books,
an imprint of Springer-Verlag New York, Inc.
A member of BertelsmannSpringer Science+Business Media GmbH

Copernicus Books
37 East 7th Street
New York, NY 10003
www.copernicusbooks.com

Library of Congress Cataloging-in-Publication Data
Fischer, Daniel.
Mission Jupiter / Daniel Fischer ; translated by Don Reneau.
   p.   cm.
Includes bibliographical references and index.
ISBN 0-387-98764-9 (alk. paper)
   1. Galileo Project.   2. Jupiter (Planet)—Exploration.   I. Title.
QB661.F57   1999
629.43'545—dc21   99-31342

Manufactured in the United States of America.
Printed on acid-free paper.
Text designed by Irmgard Lochner.
Translated by Don Reneau.

9 8 7 6 5 4 3 2 1

ISBN 0-387-98764-9   SPIN 10711962

Foreword to the Original

Edition

When Galileo Galelei discovered the four major moons of Jupiter in 1610—the Galilean moons, as we call them—he could hardly have dreamed that scarcely 400 years later a spacecraft named after him would be traveling there to give us a first-hand look at the gas giant and its satellites. Still, the Galileo project has to be regarded as one of the most spectacular undertakings in the history of unmanned space flight. Capturing such an enterprise in all its aspects between the covers of a book would seem an impossible task. It necessarily covers a quarter century of planetary research, along with the largest and most expensive interplanetary probe of its epoch making the longest journey to the biggest and most complicated planet in the Solar System—as well as the many discoveries it made on the way. Certain fundamentals have to be introduced, and interpretations of the data must be conveyed in detail, even while some of them have been understood by only a handful of experts. The project has had an eventful history, often experiencing fewer highs than lows, and in addition to telling that story, the very best images must be selected and explained from among a multitude of available photographs. Writing the book is inevitably accompanied by a nagging sense that most of the mass of information and analysis collected loses its place to something else even more important.

I have received assistance from a variety of quarters, illuminating some of the specialized aspects of this enormous scientific undertaking, even while others remained in the dark. The excitement of witnessing raw images arriving in fragments from Jupiter, shared with the public during the Voyager mission, this time was withheld. My hope, nevertheless, is that I have allowed myself to be influenced more by the relevance of particular details than by disorganization in the source base.

This book covers essentially three complex areas of subject matter, presenting them in five chapters. The first concerns the lengthy chain of events leading up to the Galileo project, which is tightly interwoven with the stories of its two predecessors, the Pioneer and Voyager probes. These early adventures in the Jovian system lie over two decades in the past now, making it more than appropriate to recall in some detail their path-breaking careers in space flight. Doing so supplies an occasion for presenting the basics about Jupiter and the system surrounding it, the same facts underlying the scientific conception of the Galileo project. The result is a survey of the lengthy development of the Galileo mission, from its now-forgotten beginnings to its repeatedly delayed launch date. From there we move through a long list of difficulties—and an equal number of triumphs—over the course of a six-year journey to the giant planet. Arrival day is presented, along with the major discoveries made by Galileo's atmospheric probe. Even now, two years after the probe's brief and intense encounter with the gas giant, scientists are still at work analyzing the complex data.

Galileo spent 1996 and 1997 observing whatever came in range of its sensors, based, of course, on an orbital trajectory calculated with the utmost precision. The nature of the chronology changes at this point, when we turn to the book's central focus on all of Galileo's discoveries during its two-year primary mission through the Jovian system. Among them are discoveries about Jupiter's four major and many minor moons, about Jupiter itself, its ring and powerful magnetosphere. The last part of the book ventures an overall view of the Jovian system, which like all scientific statements is necessarily provisional. With a parting look at the extension of the Galileo project through 1999, at the quite comparable Cassini mission to Saturn, and ideas for future Jupiter missions, the book comes to an end.

I had been planning a book on Jupiter for a long time, but 1998 seems to have been the right time to get started on it. The primary mission is now history, even though the two-year extension of the mission has since gotten under way. The steady stream of new images has let up, and it is time to strike a balance, which, incidentally, is what scientists are doing as well. By now the first in-depth examinations of the central questions have already come out in scientific journals, and in any case the thought of waiting for all the data to be analyzed is as utopian as ever. Twenty years following the Pioneer and Voyager visits to Jupiter, new analyses of those data are still being published.

The beginnings of the Galileo project lie so far in the past by now that it was often hard to locate either documents about or witnesses to the events in question. Particularly useful in reconstructing the history of the project were contemporaneous articles written by Wolfgang

Engelhardt. My chronology of the mission from 1986 to 1997 is based on countless publications in the technical journals of both the natural sciences and the space flight industry, as well as on status reports and press releases from mission managers. The latter sources, fortunately, I had already processed into a series of articles for *Sterne und Weltraum* and *Skyweek*—otherwise, it might have taken me years to subdue the mountain of documentary material. Also part of the story are scientific publications on the various discoveries of planetary objects along Galileo's way, the sometimes up-to-the-minute reports of findings at the international conferences of planetary scientists in 1992, 1996, and 1997, and the Baltimore conference on the crash of the comet Shoemaker-Levy 9 in 1995. Invaluable firsthand information came, as always, from the press conferences on Galileo's discoveries held by NASA, which I was able to follow either live via NASA's special television channel or through reports by third parties.

For their assistance I would especially like to thank Ms. Susanne Hüttemeister (at the time, of Cambridge, Massachusetts) as well as the friendly staff at the European Space Agency's headquarters in Paris, who made it possible for me to follow developments either by recording NASA television via ISDN or even, the night Galileo arrived at Jupiter, allowing me to experience the events in real time. Many interesting tangents to my primary material came from this source, and but for my encounters there, they would not have found their way into the printed literature. For a depiction of the personal dramas that are also part of the history of the Galileo project, there was simply not enough room. When the narrative turns to the failure of critical systems—the main antenna and the onboard tape recorder, for example—and to desperate efforts of scientists and engineers to save the mission, I hope readers will be able to put themselves in the place of a Galileo "tiger team" and feel the special drama of these weeks. The same, of course, also applies to the steadily increasing number of triumphs.

Thanks are due to many people who contributed individual details to this overall depiction of the Galileo mission, in the case of some during the final days the manuscript was being prepared. Deserving of special mention are Fritz Neubauer of the University of Cologne, Germany, who made himself available for an interview in the midst of an important conference, the dust researchers working with Eberhard Grün, who steered much valuable material my way, and the Washington editorial staff of *Nature*, who faxed parts of the latest publications the day before the book went to press. Very special thanks, finally, to the book's editor and publisher for sensibly allowing what was ideal for the contents of the book to determine the publication schedule.

Daniel Fischer
Königswinter, March 1998

Foreword to the English

Edition

Published almost three years after the German edition, the English version has been updated in numerous places in the "Looping from Moon to Moon" and "What the Future Holds" sections—the latter with the help of Jim Erickson, project manager until early 2001. New scientific insights were added that had become available from recent publications and through picture and press releases, and about 25 interesting black and white images were included, without sacrificing earlier material. The very latest developments of 1999, 2000, and early 2001 are being presented as progress reports at the end of several sub-chapters. While few of the earlier conclusions have been overturned completely by the continuing analysis so far, some mysteries have deepened. At the same time, other aspects of the Jovian system have become surprisingly clear. Since the middle of 2000, for example, the existence of an ocean of a salty liquid under the icy crust of the moon Europa can be considered proven, thanks to Galileo's magnetometer. As this edition goes to press in the year 2001, Galileo is still alive and well and has just completed yet another mission milestone: joint observations of Jupiter's vast magnetosphere, together with that other outer planets explorer Cassini, which is on its way to Saturn. And if no major electronic failure kills the spacecraft, its mission will continue well into 2003, to be ended in a controlled fashion by sending Galileo right into Jupiter's atmosphere. Yet more discoveries can be expected when the orbiter makes some more close flybys of the moons Callisto, Io, and perhaps Amalthea during its final orbits, as its epic journey draws to a close. Galileo has already changed mankind's view of the largest planet of the Solar System and its companions forever. Enjoy the trip!

Daniel Fischer
Königswinter, March 2001

Contents

Chapter 3

Arrival and the Atmospheric Probe   113

Chapter 4

Looping from Moon to Moon   143

Chapter 5

What the Future Holds in Store   263

Chapter 1

The Long Journey to Jupiter

December 7, 1995: Rendez-vous with a Giant

It is December 7, 3:09 P.M. in California. "They continue to look worried," says a NASA spokesperson. The tension is palpable. "They" are the flight control specialists at JPL, the Jet Propulsion Laboratory on the outskirts of Los Angeles. For hours they have had their gigantic antennas trained on Jupiter, waiting in vain for a signal. Nearly a billion kilometers separate Earth from Jupiter, but the signal should have come long ago. Being anxiously awaited was a single bit of information that would announce the beginning of an extraordinary mission. At this moment, Jupiter is about to receive Galileo, a visitor from Earth. It is the most expensive and most complicated interplanetary probe ever to have been sent to the outer solar system, and though not the only spacecraft to have traveled this distance, it certainly is the first designed to spend years exploring the giant planet.

Nearby worlds—the Moon, Venus, and Mars—have had a number of Earth emissaries over the 30-odd years of the space program. Spacecraft have been put into orbit around planets; they have probed their atmospheres and even made a soft landing on the surface. Humans have actually set foot on the Moon. Yet, until today, the systematic exploration of the distant planets beyond the Martian orbit was only a dream. A few probes had made the journey, traveling for years to spend only a few days at their destination. No one had known how to supply a spacecraft with the fuel reserves needed to maneuver into orbit around Jupiter, Saturn, Uranus, or Neptune. Without adequate fuel reserves, the spacecraft had no way of slowing down. Nevertheless, the "encounters," as the early flyby visits were called, yielded literal torrents of images and data for the scientists. Shooting past Jupiter in the 1970s, the two Pioneer and the two Voyager probes rewrote the textbooks.

Launching Galileo meant that the time had finally come for taking a systemic look at the biggest planet in our Solar System. A spacecraft jam-packed with scientific instrumentation would be going into orbit around the planet. And for the first time, an equally well-equipped atmospheric probe would descend into the outer region of the gas giant, sending back data before finally being vaporized by the planet. These are the events for which scientists are poised on the afternoon of December 7, 1995. They are waiting anxiously for confirmation that the probe is in working order, transmitting data to the Galileo orbiter. In two hours, assuming that all goes as planned, Galileo will enter into orbit around Jupiter. But for now, everyone waits for the first, all-important signal.

The mood at JPL that afternoon recalled a day 25 years ago, when the damaged Apollo 13 capsule had burned the last of its fuel in a desperate attempt to get back to Earth, and the transmission signaling that the three astronauts had survived reentry came several minutes late. The final suspenseful moments of the *Apollo 13* film are in the mind of more than one onlooker, whether in California or in Europe at the Paris headquarters of the European Space Agency, where scientists are also monitoring the events. NASA is well acquainted with what it takes to establish such milestones in the history of space flight. Mission control at JPL, where commands are formulated and transmitted to Galileo and incoming information is received, has been hermetically sealed off. Included in the information is the spacecraft telemetry, data regarding the status of the individual subsystems it carries onboard. Here in the Mission Support Area (MSA), only those with critical contributions to make are allowed in. Even the biggest names in space journalism have to wait outside.

The public gets to share in the suspense and, hopefully, success of the mission thanks to a camera technicians have installed in the MSA. Every word goes out on a live feed to NASA Select TV, the space agency's own television channel. In a small television studio set up in a neighboring room, veterans of the first Jupiter flyby in the 1970s try to explain what is going on. The broadcast will last several hours, made available not just to American television stations, but to basically everyone in North America via satellite and cable.

Unfortunately for Rich Terrile, anchor of this all-night broadcast, the immediate task is explaining what could have happened to delay word that Galileo had arrived intact. It seems to be a repeat of a problem that had threatened not long ago to bring the entire mission to an end. The large high-gain antenna of the space probe had never opened correctly, consigning all communication needs to a relatively tiny low-gain antenna. The stream of telemetry coming from Jupiter is too thin, only a few bits per second making the 52-minute trip to show up directly on the computer screens—on which all eyes in the MSA are nonetheless riveted. NASA's Deep Space Network, which has the job of maintaining contact with all space probes in the Solar System, is picking up what signal there is. Through giant parabolic antennas, the largest spanning as much as 70 meters, a handful of experts listen in at three stations spread across the globe, in California, Spain, and Australia.

At this moment, both Jupiter and Galileo lie beyond the horizon for two of the tracking stations, in Goldstone, California, and Canberra, Australia. The data flow is so thin that information must be allowed to gather into packets before being sent on to JPL, where the tension has become nearly unbearable. The atmospheric probe separated from the Galileo orbiter a full six months ago, after all, and until now there has been no way to run checks on its functions. Following comprehensive testing, first in the laboratory and then during the long trip to Jupiter, when the probe and the orbiter were still connected, there has been nothing but mathematical models to rely on. And one thing could scarcely be more obvious: this is the only chance the probe will have to execute its precisely calculated approach to the Jovian atmosphere.

Sixty-five minutes have passed since the probe began encountering the denser regions of Jupiter's atmosphere, 63 minutes since the planned onset of data transmission. And Jupiter is only 52 light-minutes (or radiominutes) away. Atmospheric scientist Andy Ingersoll, Rich Terrile's expert source in the studio, urges everyone to stay calm. The very same thing happened ten years ago, he says, when scientists had to wait for the first signals to arrive from balloons descending into the atmosphere of Venus.

Suddenly, at 3:11 P.M., contact! Jubilation breaks out in mission support. A single bit of information has confirmed that the Galileo orbiter is receiving signals from the probe, now in the depths of Jupiter's

Why is Jupiter one of the most fascinating planets?

- It is the largest of all the planets, containing 70 percent of their combined mass.
- It has many moons.
- It has an unusual ring system.
- Its moon Io is the most active volcanic body in the solar system.
- It produces the largest magnetosphere.

Jupiter by the numbers:

Mean diameter:	142,796 km	= 11.2 times Earth
Mass:	1.9×10^{27} km	= 319 times Earth
Density:	1.33 g/cm^3	= 0.24 times Earth
Equatorial tilt:	3.1°; Earth	= 23.45°
Mean distance from the Sun:	778,631,330 km	= 5.20 AU

(AU = the mean distance between Earth and the Sun = 149.6 million km)

Minimum distance from Earth:	3.93 AU
Maximum distance from Earth:	6.46 AU
Orbit eccentricity:	0.048
Inclination of orbit to ecliptic:	1.305°

Orbital period:
  sidereal (in space, relative to stars): 4,335.22 days = 11.9 years;
  synodic (same position relative to the Sun, from Earth): 398.9 days

Jupiter's moons:

Name	Mean diameter (km)	Distance from Jupiter (1,000 km)	Orbital period (sidereal, days)
Metis	40	128	0.20
Adrastea	20	129	0.30
Amalthea	188	181	0.50
Thebe	100	222	0.67
Io	3,630	422	1.77
Europa	3,138	671	3.55
Ganymede	5,262	1,070	7.15
Callisto	4,800	1,883	16.7
Leda	16	11,094	238.7
Himalia	186	11,480	250.6
Lysithea	36	11,720	259.2
Elara	76	11,737	259.7
Ananke	30	21,200	631 retrograde
Carme	40	22,600	692 retrograde
Pasiphae	50	23,500	755 retrograde
Sinope	36	23,700	758 retrograde

Twelve more small moons, still nameless, were discovered in 2000.

atmosphere. The applause is deafening in JPL's Van Karman Auditorium as well, where hundreds of project scientists and fans have gathered. Program managers are falling into each other's arms. It has been 74 months since liftoff, and finally the climax of the Galileo mission has begun. A sequence of events is under way that will go on for several years, fundamentally altering science. It is the dawning of a new age in planetary research.

The King of Planets Has Always Beckoned

What is the brightest object in the night sky? The Moon, of course, has no rivals in size or brightness. But after that comes our neighboring planet Venus, also called the Morning (and Evening) Star, and the third brightest object is Jupiter. Often it is the brightest of all. Venus and especially the Moon change their positions among the stars very quickly, from night to night. Jupiter, reflecting pure white light, alters its position only slowly, over the course of months. No wonder the ancients accorded Jupiter a lead role among the planets, which they understood to be "wandering stars." Jupiter represented the supreme godhead; in Greece, he was Zeus, and he was Optimus Maximus or Jupiter in Rome. *Iovis*, from the Latin, lives on in technical astronomical expressions such as "Jovian," in the sense of something belonging or pertaining to Jupiter. Jupiter alone ruled the heavens, the cause of all motion, including weather phenomena such as rain, hail, and thunder. After a storm, it was Jupiter who drove away the clouds and restored order. It was only fitting that the planet moving most evenly through the signs of the zodiac, visiting each one in an approximate annual rhythm, would be Jupiter.

Once the Sun was recognized as the center of our Solar System, Jupiter's routines became easy to understand. The planet is on average

5.2 times as far from the Sun as Earth, taking almost 12 times as long to make a revolution. A year on Jupiter lasts 11.86 Earth years. For us, all this explains why Jupiter takes up a position directly opposite the Sun every 399 days, which causes it to remain visible the entire night through. Astronomers term that "opposition." When the Sun goes down, the planet comes up. It reaches its highest point in the southern sky around midnight, and it sets again as the Sun rises.

Jupiter's orbit, strictly speaking, has a slightly elliptical shape. The average distance separating Jupiter from the Sun is 778 million kilometers, but that figure varies by as much as 38 million kilometers. Of course, it was only when astrometry, the science of astronomical measurement, began in the sixteenth century that these things were learned about Jupiter. Not long after that the telescope was invented, astronomers were able to see that Jupiter is the largest planet in our Solar System, second in size only to the Sun. With an equatorial diameter of 143,000 kilometers, Jupiter is 11 times larger than Earth, although it is also a relatively "flattened" planet—a mere 133,000 kilometers pole to pole.

Jupiter is also immense in volume, about 1,320 times the volume of Earth. Jupiter's mass, however, exceeds Earth's by only 319 times, putting the planet's density at 1.33 grams per cubic centimeter, scarcely any denser than water. And still, the gas giant is so massive that it influences the operation of the Solar System. It has 1/1,047 the mass of the Sun, which means that the center of gravity in the Jupiter-Sun system lies just outside the boundaries of the Sun itself. Through a simple pair of binoculars, Jupiter is visible as a small disk in the sky, despite how far away it is, which can be as much as a billion kilometers. A small telescope is enough to show the planet's distinct flattening (1/15). Jupiter's brilliance in the night sky comes from the high reflectivity of its "surface"—it took scientists a very long time to determine that the planet is made up entirely of clouds, with nothing solid underneath. Still, it reflects back an unbelievable 44 percent of light shining on it. The face of the planet appears a dull yellow color through a telescope with no color filtering. Jupiter is radiantly white to the unaided eye, however, an effect caused by the high level of contrast between the bright disk that the eye registers as a shining point and the dark background sky.

Many of the photographs of Jupiter's clouds in this book show a color-enhanced version of the planet, and often they have been composed out of a number of individual exposures in specific color ranges. The effect is to bring out details that could not be seen otherwise, including the way the planet is built up of layers. There are, however, occasional shots of Jupiter in its natural dull yellow and brown tones, as well. Some experienced amateur astronomers have managed to do an "image analysis" like this in their minds. It takes a great deal of looking through large telescopes, but they learn to perceive fine nuances of color that are invisible to the occasional observer. Whoever might be inspired by this book to visit a public observatory should be warned that the first glimpse of the "king of the planets" is often disappointing. However, patience will be rewarded. What at first seems to be no more than a yellowish oval in the eyepiece soon starts seeming to have two darkish bands stretching parallel to the equator, above and below it. If the air is clear (and the telescope is good), more and more of the details become visible: more stripes, and also holes and dark spots inside and between the bands.

Within minutes, it becomes possible to relive the history of Jupiter exploration in fast forward, and it often happens that those who do never escape the spell of this special area of planetary research. The planet's rapid rotation, for example (which also causes the flattening), becomes noticeable after only a few minutes in the quick movement across the surface of cloud details. It takes just under ten hours, one day on Jupiter, for the clouds to complete a rotation.

With a pair of well-supported binoculars or a small telescope, the first object to appear is, of course, the flattened disk. Then, within a few hours, nearby points of light will have changed their position noticeably relative to Jupiter. At that point, observers are seeing what both Galileo Galilei and the German mathematician Simon Marius discovered in January of 1610 with the recently invented telescope: the dance of four of Jupiter's moons. These four are today also called the Galilean moons, although their individual names—Io, Europa, Ganymede, and Callisto—actually came from Marius. Who could have dreamed back then that each individual moon would present us with a completely different world, making Jupiter and what surrounds it practically its own small planetary system?

The First Pioneers

March 2, 1972: on this day the first-ever interplanetary space probe destined to explore the outer Solar System is being launched. Although Pioneer 10 weighs only 260 kilograms, it is almost all an Atlas-Centaur rocket can do to send it hurtling toward Jupiter. The trip will take only 21 months, but it is a journey into the unknown. In between Mars, which has by then already been the target of numerous space probes, and Jupiter lies the Asteroid Belt. A good bit of mission planning has been taken up by the Asteroid Belt, where literally millions of "minor planets" revolve around the Sun, having been prevented by the pull of Jupiter's gravity from condensing into another normal planet. In science fiction, the Belt is often being imagined as gigantic boulders coursing closely to each other through space—and 25 years ago, astronomers did in fact not know much better. However, the likelihood of a collision with a giant boulder was basically nil. Scientists had gathered at a conference on asteroids the previous year and had reassured themselves on that point. The real

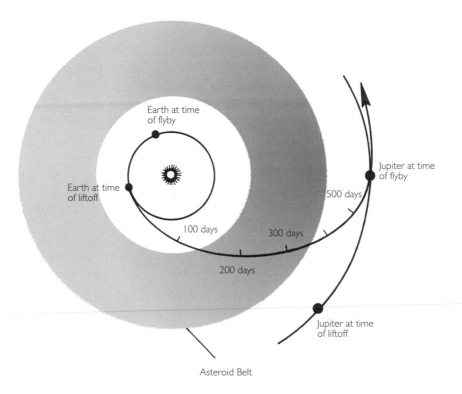

Earth at time
of flyby

Earth at time
of liftoff

Jupiter at time
of flyby

100 days

200 days

300 days

500 days

Jupiter at time
of liftoff

Asteroid Belt

A direct route to Jupiter: a probe as small as Pioneer 10 can get there quite quickly. The trip at the time of the Pioneer probes lasted less than two years, even with an early-1970s rocket.

unknown was dust in between the boulders, which was being produced constantly by collisions between the boulders. At the high speed the spacecraft would be racing away from Earth toward the outer Solar System, a collision with even a tiny dust particle could spell disaster.

One of the objectives of the Pioneer project—both Pioneer 10 and its sister probe, Pioneer 11, which lifted off several weeks later on April 15, 1973—was to prove that the Asteroid Belt was passable. And there was more concern yet about the intense radiation the probes would encounter in the immediate vicinity of Jupiter itself. The electronics of the probes would require special shielding, because of the charged particles that become trapped in such large numbers in the planet's powerful magnetic field. The Pioneers got their name for a good reason. They would prepare the way for a generation of much more sophisticated space probes that would journey out beyond Jupiter to Saturn, Uranus, and Neptune. The favorable alignment of the four planets in relation to each other in the 1970s and 1980s would make it possible to accelerate probes from one planet to the next, sending them on the longest journeys ever undertaken from Earth. The "grand tour" to be made by the Pioneer probes was mainly a scouting mission, and they were deliberately kept simple. With a total cost of $100 million between them, their onboard scientific payload was limited to only 30 kilograms. They had dust detectors and telescopes for larger objects that happened to be flying by, instruments to measure magnetic fields, as well as other detectors for high-energy radioactive particles and for Jupiter's own radiation, which ranges from ultraviolet to radio waves.

Neither Pioneer had a camera in the narrow sense, but they did carry a kind of light meter capable of scanning Jupiter's clouds. The raw color images they produced were blurred and distorted, but after a lot of image processing—not easy with the computers of the early 1970s—they were much sharper than the best photographs that could be taken from Earth. Just arriving at the destination and sending a signal back home from 45 light-minutes away was an accomplishment. Pioneer 10's transmitter had only 8 watts of power, so the energy reaching Earth, despite its 3-meter parabolic antenna, was nearly infinitesimal. For that amount of energy to accumulate enough to light a normal candle for a thousandth of a second it would take 20 million years.

Flying through the Asteroid Belt turned out to be harmless child's play, but the magnetic field near Jupiter was even more dynamic than scientists had anticipated. On November 26, 1973—still a distance equal to 54 Jupiter diameters away from the planet (almost 8 million kilometers)—Pioneer 10 went through "bow shock," passing through the point where the solar wind—a constant stream of charged particles traveling at supersonic speed—encounters the resistance of Jupiter's powerful magnetic field. Over the next few days, especially strong solar winds pushed the edge of the magnetosphere back toward Jupiter, and once again Pioneer 10 was engulfed in pure solar wind. The probe had to plow through tremendous turbulence, going through bow shock for the final time 25 Jupiter diameters from its destination. From that point, for a few days, it became part of the Jovian system, inaugurating the truly exciting part of the trip, which would climax in a Jupiter flyby on December 3, 1973, only 131,000 kilometers above the clouds. Would Pioneer's electronics survive the intensifying bombardment of charged particles in Jupiter's magnetic field? The onboard computer did execute a few unplanned commands because of the particle assault, triggering an automatic shutdown, but the point had been made. It was possible to explore Jupiter from close up.

And what there was to explore! The axis of Jupiter's magnetic field is tilted 15° relative to its rotational axis, so the field does not pass exactly through the center of the planet, and the polarity of Jupiter's field is reversed, compared to Earth's. The big surprise, however, lay elsewhere: the charged particles trapped in the magnetosphere are heavily concentrated in the plane of the *magnetic* equator. That means, because of the tilt between the magnetic axis and the rotational axis, the disk containing the greatest number of particles "wobbles" in rhythm with Jupiter's rotation. Pioneer was able to register variations in the intensity of Jupiter's radiation over a ten-hour period, making it possible for the first time to put numbers on radioactive exposure and develop a shielding strategy for future space probes.

Pioneer 10 also made fundamental discoveries about Jupiter itself, or more precisely, its atmosphere. It was able to pick up traces of helium at a wavelength in the far-ultraviolet (58 nanometers), and in the ultraviolet (122 nanometers) it detected the glow of hydrogen. Long an objective of planetary research, measuring the abundance of the two most elemental gases near the outer planets carried implications far beyond the study of the Solar System, reaching all the way to cosmology. Neither the Pioneers nor their successor probes, the Voyagers, were expected to obtain definitive answers in these matters. That was a task scientists always knew would be reserved for Galileo's atmospheric probe two decades later. Nevertheless, Pioneer 10's infrared instrumentation fundamentally advanced knowledge about Jupiter's internal nature. For one thing, the planet gives off two-and-a-half times as much heat as it receives from the Sun! Under the force of its own gravity, Jupiter continues shrinking even today, releasing the potential energy involved in the form of heat.

The Pioneers also made new discoveries about the mass and density of Jupiter's moons. Io has a mean density of 3.5 grams per cubic centimeter, for Europa the figure is 3.0, and then come Ganymede and Callisto with 1.9 and 1.7, respectively. Density, in other words, decreases with increasing distance from Jupiter. Pioneer learned that Io, with its weak ionosphere, probably does have an atmosphere, even if one amounting to a mere 1/100,000 of a millibar of surface pressure. But there were no spectacular photographs, only blurred images of a couple of bright and dark spots on Ganymede. With a resolution no smaller than 400 kilometers, the pictures offered no hint of what strange worlds would be revealed by photographs showing 1,000 or 10,000 times the detail. Pioneer 10's photographs of Jupiter itself, however, did just that—without even approaching the quality of shots taken by Pioneer 11 during its closet approach to the planet on December 3, 1974. It came within 42,000 kilometers of Jupiter's clouds, ten times closer than Pioneer 10 had ventured, and Pioneer 11 came through the radiation bath intact.

The trick lay in Pioneer 11's custom flight path. Shooting through Jupiter's equatorial plane at a 50° south-to-north inclination set an

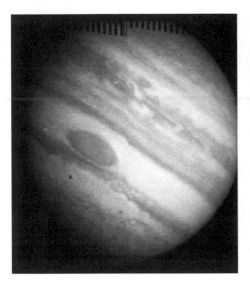

In views of the entire planet like this Pioneer photograph from the early 1970s, the Great Red Spot made a particularly dramatic impression.

Due to its unusual orbit, so far only Pioneer 11 has been able to examine Jupiter's polar regions.

upper limit on the overall dose it received. The special trajectory sent the space probe cutting directly across the Solar System toward Saturn, where it arrived, still functioning, five years later. And it set up the scanner camera for some unusual perspectives on the cloud formations near Jupiter's poles, which can scarcely be seen from Earth and have not been explored by any space probe since. Scientists had made progress interpreting the cloud structures in the year since the first Pioneer mission, and now they were getting the clearest pictures ever. Key to the formations seemed to be thermal sources inside Jupiter. Seeking to escape, the heat sets up tremendous convection flows. The *light* zones encircling Jupiter are composed of rising streams of gas, which in the *darker* zones are going back down. Infrared measurements from Pioneer 10 had shown that the uppermost areas of the

Even the early Pioneer probes were capable of resolving the fascinating swirls in Jupiter's clouds this sharply, better than any ground-based telescope. They opened a new window in planetary research.

Pioneer 10 captured this image of Jupiter's Great Red Spot and the shadow of the moon Io on December 2, 1973, from 1.5 million kilometers away.

light zones are cooler than the dark stripes by about 9° Celsius, which means that they are about 20 kilometers higher than the stripes.

Pioneer 10 and 11 each made a special study of Jupiter's famous Great Red Spot, a gigantic oval storm system that has been known as part of Jupiter's atmosphere for 300 years. The oval stretches 30,000 kilometers in length, with current measurements putting it about 8 kilometers higher than the surrounding clouds. Photographs from Pioneer 11 showed clearly for the first time how the storm is embedded in more general currents north and south of it, influencing them at the same time—and they proved that the storm as a whole is rotating in a counterclockwise direction. Pioneer images also brought other smaller storms into relief, as well as complex swirl patterns on the edges of faster moving currents, which had been predicted on Earth in advance of the mission. On its flyby, Pioneer 11 repeated the intensive examination of the magnetic field (this time using two different magnetometers at once and getting good agreement in the results) and Jupiter's particle radiation. Using additional data from the Pioneer 10 mission, astronomers were able to calculate the intensity of the magnetic field at the altitude of the visible clouds. At approximately 4 gauss, it is ten times more powerful there than on the surface of the Earth.

Thanks to improved detectors, Pioneer 11 was able to perform a more precise analysis of Jupiter's radiation belt. Image analysts could clearly distinguish the outlines of structures inside the inner radiation belt, apparently the result of the way Jupiter's orbiting moons are constantly "stirring" the magnetic atmosphere. The effect is most noticeable on Io, where the number of energetic protons (hydrogen nuclei) shrinks by a factor of 70 near the surface. But Pioneer 11 found a lot more happening on Io. The ultraviolet spectrometer discovered a gigantic cloud of hydrogen atoms surrounding the moon in a broad area 120° along its orbit. Nothing of that sort was identified on either Ganymede or Callisto. Io's hydrogen ring could be explained only by the presence of a thin atmosphere, which would have to be replenished constantly by gases escaping from inside the moon. That was the first indication that Io was something more than a dead clump of ice. Still, no one could have guessed just how lively the Voyager probes and Galileo would prove it to be.

The Voyagers' "Grand Tour"

To visit *all* of the outer planets in the Solar System—Jupiter, Saturn, Uranus, Neptune, and even Pluto—this has been the dream of planetary scientists since the 1960s, a few years after interplanetary space flight first got under way with Mariner 2's Venus encounter in 1962

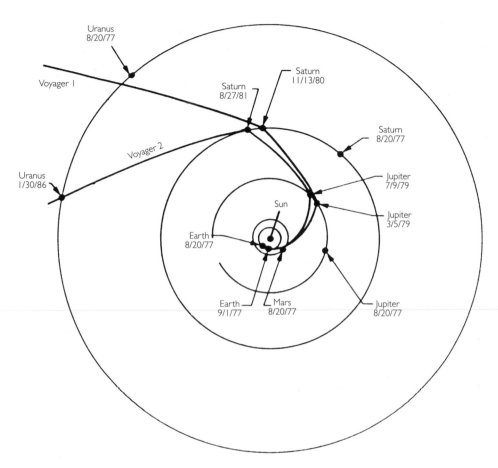

The grand tour by the Voyager probes to Jupiter and Saturn (and for Voyager 2, on to Uranus and Neptune).

and then the arrival of a probe at Mars. The optimism was based on the favorable planetary alignment mentioned in the previous section. Coming only once every 175 years, this rare configuration would make it possible for a single spacecraft to visit several planets in a relatively brief period of time. The assumption had always been that this planet surfing would begin with a Jupiter flyby, followed, for example, by Saturn and Pluto, or by Saturn, Uranus, and Neptune. The only impossible route was the one including all five planets, one after the other. Flight controllers always used the same trick. They steered the probe past a planet in exactly the position that allowed it to steal a tiny portion of the planet's momentum, picking up a "gravity assist" to speed it with extra force along its way without consuming any valuable fuel in the process. This fascinating idea was named the "Outer Planets Grand Tour" (OPGT), and NASA experts began working out the details in 1969, the year of the first Moon landing. The Pioneer flybys themselves had won approval as part of this effort. "We maintain that a thorough investigation of the outer parts of the Solar system is one of the most important tasks of extraterrestrial exploration in this decade." The National Academy of Sciences of the U.S. stated to the OPGT in 1971. "We recommend the development of space probes . . . and a series of four launches in the late 1970s."

By now the enthusiasm of the Apollo years had already waned. The $750 million it would cost to build, launch, and operate the four space probes envisioned for the mission was no longer available. In 1972, NASA had no choice but to give up the grand tour in an effort to save at least the trip to Jupiter, with an accompanying swing toward Saturn. The number of probes was cut in half to two, and they were to be constructed much more economically, now requiring them to last only a couple of years. That both of them would still be in the best of operating health 20 years later was at most someone's secret dream on July 1, 1972, when the project "Mariner-Jupiter/Saturn 1977" (MJS77) officially got under way. This unwieldy initial designation was later changed to Voyager. Costs ultimately climbed to $350 million for both probes, in addition to $72 million for the rockets and $33 million for flight control and data analysis.

More than a thousand engineers, scientists, and managers worked at the Jet Propulsion Laboratory and in the industry to have both Voyagers ready for their scheduled liftoff date in the summer of 1977. Identical to each other, the probes also took similar routes to Jupiter, at which point their assignments became more specific. If something had happened to one of them on liftoff, or if one had been lost under way, the other could still have carried out the majority of the scientific experiments. This wise double-launch strategy had proved itself in the 1960s (when more than one interplanetary probe had been destroyed on liftoff) and, unfortunately, would be called upon one last time for the Voyager probes. Each Voyager had a mass of 825 kilograms (2.1 tons fully fueled), of which 105 kilograms were devoted to scientific experiments. The probes were improved versions of the Mariner series, which had already demonstrated its longevity in the inner solar system. As had already been the case for the Pioneers and would come into play again for Galileo and the Cassini Saturn orbiter, the only way to meet energy needs required using radioisotope batteries. Radioactive plutonium produces heat when it decays naturally, and the heat is transformed into electrical energy.

The "grand tour" began on August 20, 1977, with the launch of Voyager 2. Voyager 1 followed on September 5, taking a different route that would get it to Jupiter faster. The rockets being used at the time were Titans with Centaur upper stages, but it would be the last time for 15 years that interplanetary probes would be sent on their way by "one-way" rockets. From now on, the space shuttle would take over transport duties, ultimately putting three probes into space. But this strategy did not pay off. Since 1992, the US has again been using Titan and Delta rockets to launch planetary probes, because of their combination of considerable cost savings and enhanced flexibility. Technologically, the Voyagers were much superior to the Pioneers, which unfortunately includes the disadvantage that more can go wrong. The liftoff of Voyager 2 went perfectly, and the probe was right on track for Jupiter. But, almost from the first minute, mysterious disturbances had flight controllers holding their breath. Serious malfunctions, threatening major aspects of the mission, followed repeatedly in the next few months. For a time, the camera platform on Voyager 1 was jammed. The main radio receiver on Voyager 2 stopped functioning altogether, and not all parts of the backup system were working. That Voyager 2 would go on to become a great success story in the exploration of the distant reaches of the Solar System speaks volumes about its robust design—and about the ingenuity of its human operators.

The excitement began in early 1979. Voyager 1 had arrived within 50 million kilometers of Jupiter, delivering over 500 pictures under way, among them film sequences of clouds in motion. Noticeable changes had occurred since the Pioneer years of 1973 and 1974. The Great Red Spot in particular had lost much of its color (something that has occurred frequently in the past). At first glance, it might have seemed that this was a completely different planet. By early February, Jupiter was taking up the onboard telecamera's entire field of view. Several of the ten other Voyager instruments began operating about this time as well, and all together they would mount a systematic exploration of the planet. There were infrared and ultraviolet spectrometers, receivers for radio and plasma waves, magnetometers, and a variety of charged-particle detectors. The Voyagers were classic examples of a certain class of large space probes. They transported an arsenal of the best available instruments to a distant planet, aiming to uncover all of its secrets. Faced with a cost explosion and tight budgets in the 1980s, NASA withdrew from this commitment, which the Russians had long been attempting to match. And Galileo and Cassini will be the last probes of this sort for a long time to come. Whether the new quick and cheap approach to planetary exploration (of which the 1997 Mars Pathfinder is a sterling example) will deliver the same high-quality scientific information over the long run as the "dinosaurs" of the 1970s remains a matter of dispute among US scientists.

In February 1979, Voyager 1 crossed through the orbit of Sinope, the outermost moon of Jupiter—which remained a long 23 million kilometers away. Already, photographs were showing much more detail than analysts had been able to tease out of the scanned images of the Pioneer probes. Instruments had begun picking up the first radio signals from Jupiter, which increased in intensity whenever one of Jupiter's polar regions was turned toward Voyager. Scientists were soon celebrating other discoveries as well. Voyager was registering extremely low frequency radio waves, with wavelengths in the dozens of kilometers. The waves had gone undetected until this time, because they cannot penetrate Earth's atmosphere. The source of the radio waves seems to be not Jupiter itself, but instead the magnetosphere surrounding the planet, perhaps in connection with the dense plasma torus produced by the moon Io. There was also a big surprise in the ultraviolet. At this short wavelength, Jupiter's spectacular polar lights even radiated across the daytime side of the planet.

Activity picked up at the end of February in preparation for the closest approach to the planet on March 5, 1979. A new image was coming in every 48 seconds now, revealing in ever increasing detail the structure of Jupiter's clouds. In the hours following Voyager's "close encounter," it would pass by each of Jupiter's largest moons, taking photographs. But first, Voyager 1 had to enter the magnetosphere—and in making that landmark passage, the spacecraft was late. Solar activity had picked up since 1974, with the increased force of the solar wind pushing the point of bow shock closer to Jupiter. It was not until February 28, still 43 Jupiter diameters away from the planet, that the probe passed through the boundary for the first time. But the boundary itself kept moving back and forth. Now only 22 diameters away, Voyager 1 finally passed completely into Jupiter's powerful magnetic field on March 3, 1979. Already, JPL specialists had turned the constant stream of photographs of the planet into lengthy time-lapse films, upsetting a number of established ideas about currents in the atmosphere, including some of the Pioneer findings. The currents surrounding the Great Red

Jupiter's Great Red Spot as seen by Voyager 1 from 9.2 million kilometers on February 25, 1979—even then details were resolved down to 160 kilometers.

Spot turned out to be especially beautiful. It took six days for a typical cloud on the edge of the Spot to move completely around it. And the rotational direction of the great swirl seemed possibly to be an anticyclone—a high-pressure zone. Knowing that storms on Earth occur in regions of low pressure, atmospheric scientists were left equally confused and intrigued.

Not only, however, were there spectacular color pictures of Jupiter for everyone to look at—not to mention the streams of data from a variety of instruments on field and particle phenomena—now there was also something *to hear.* The phenomenon had nothing to do with actual sound waves, of course, which cannot exist in the virtual emptiness of interplanetary space, but instead with plasma waves. The frequencies of these curious and complex wave phenomena—located in the charged particles swarming through the powerful magnetic field around Jupiter—happen to occur in the spectrum of sound waves that is audible to the human ear. Plasma wave scientists, finding no reason not to make use of the coincidence, quickly fed the data from their instruments into an amplifier—to the complete bewilderment of an auditorium full of journalists gathered at JPL. The "sound" made by the flow of protons Voyager was passing through, along with the background noise of the space probe itself, was something like a combination of whales singing, a snowstorm, and a car race. It was not long before people were buying recordings of the sounds to listen to at home. There was also more news about the ever-active Io. Voyager's ultraviolet spectrometer had detected sulphur ions that were missing two electrons, and the radiation from these ions at short wavelengths was astonishingly bright. The plasma ring was extraordinarily dense. And the sulfur ions were still swarming through space even ten times farther away from Jupiter, where Voyager was able to detect them directly. What, except for Io, could possibly be the source?

Yet this increasingly complex plasma phenomenon now took a back seat to the ever sharper pictures of the big Jovian moons that were now arriving. What ground-based observers (as well as the Pioneer probes) had seen as tiny disks in the distant sky suddenly transformed, almost from one day to the next, into full-blown planetary bodies. And what bodies they were! "Like Christmas" is how the planetary geologists were feeling on the afternoon of March 4. "Tonight we will begin exploring four new worlds." And another historic event had passed almost unnoticed that same morning. Speeding through

Jupiter's equatorial plane, Voyager 1 had opened the shutter on the telecamera for an 11-minute exposure, presumably of nothing. The idea, just in case Jupiter had rings, was that this was the best time to get a picture. And as a matter of fact, there was a diffuse brightness visible in the unprocessed camera image, which, like all the rest, was being fed live to a number of television monitors. Still, it took 3 days of analyzing the images to confirm that Voyager had indeed discovered a ring around Jupiter. Nothing, perhaps, compared with Saturn's spectacular ring system, but still an important finding.

March 5, 1979, the day of Voyager's closest approach, became an official occasion at JPL, when the probe, racing by at nearly 100,000

Io in front of Jupiter. This early picture taken by Voyager 1 on March 2, 1979, showed greater detail on the innermost Galilean moon than had ever been seen before. Image analysts still did not know at this time whether the brightly outlined round structures with dark centers were impact craters or volcanos.

kilometers per hour, came within 780,000 kilometers of Jupiter. The lab found itself playing host to a large number of distinguished visitors who had arrived during the night. A special television screen had been set up in the White House for the President and his family. When the close-ups of Jupiter's clouds started coming in, however, scientists, officials, and journalists alike just stood and stared at all the bizarre details in the complex pattern of swirls. Then, with celebrants at JPL already toasting the success of the mission, Voyager disappeared for two hours behind Jupiter while at JPL the success of the mission was already being celebrated as the "most exciting, fascinating and, perhaps as will eventually turn out, the most scientifically productive in the unmanned space program," as Voyager's cameraman Brad Smith was glad to say. A stunning photograph of Io had been composed from data that had come in the previous night, prompting one project geologist to liken it spontaneously to a pizza. The Jovian moon would never get rid of the nickname, even if it turned out not to be quite as bright yellow, orange, and red as it first appeared in the photo.

There were no plans for any close approach to Europa for Voyager 1, the major exploration of that moon having been reserved for Voyager 2 in July. But the first Voyager did pass within 115,000 kilometers of Ganymede, and it approached within 126,000 kilometers of Callisto. Both of these moons were full of the sort of impact craters of which there had been no sign whatever on Io, even in the highest resolution pictures. Scientists were aware that an as-yet-unknown process had to be continually erasing the evidence on Io of the ongoing impact of comets. Was Io perhaps volcanically active? There were structures that looked suspiciously like lava flows. And an article had just appeared in a technical journal, predicting "widespread and recurrent surface volcanism" of Io. A network of long fault lines on Ganymede testified to geological activity on that moon as well, albeit in the distant past. Such discoveries, nevertheless, gave geologists in particular

The first discovery of an active volcano on Io. This is the historical photograph made by Voyager 1 on March 8, 1979, from 4.5 million kilometers.

the feeling, when Voyager 1's flyby ended on March 8, that they had been introduced to an entirely new planetary system. Head scientist Ed Stone: "I think we have had almost a decade's worth of discovery in this two-week period."

One of the reasons for Stone's enthusiasm was what the probe had observed on the moon Callisto. It clearly had the oldest surface of all the Jovian moons, with far and away the most craters. Among the craters was one gigantic specimen, 3,000 kilometers in diameter, which has no rival in the entire Solar System. Then came Ganymede, with craters of its own, but in addition with a multitude of peculiar groove formations. And then, Io. With a surface less like our own Moon than some volcanic formation in Yellowstone National Park, Io looked pretty much as expected—meaning, however, only that the greatest discovery was yet to come. By March 8, with live broadcasts from the Jovian system a thing of the past, Voyager 1 shot one last photograph of Io. It was a relatively long exposure against an oblique back light, taken as a navigational photo, including a couple of recognizable stars. Io, by now 4.5 million kilometers away, appeared as a narrow sickle. But there was a second, smaller and fainter sickle just alongside. It could not possibly be a new moon, which would already have been discovered long ago. Then Linda Morabito, a project engineer

Our Picture of the Jovian System at the Time of the Voyager Missions

On Jupiter . . .

. . . the large and small cloud structures move at the same speed. That tells us that what we observe racing around the planet is actual material and not a wave phenomenon taking place inside a stationary gas.

. . . clearly defined east–west winds extend into regions near the poles. Scientists had expected to find dominant up-and-down movement in these areas, caused by convection.

. . . the atmospheric turbulence really is driven by these east–west currents, rather than by vertical convection forces, as it had seemed following the Pioneer missions.

. . . the real energy behind the cloud movements comes from the depths of the atmosphere, while the visible structures only point to it.

. . . the motion of the clouds around the edge of the Great Red Spot is anticyclonic, and it takes about six days for the clouds to move around it once.

. . . the smaller "spots" are related meteorologically to the Great Red Spot; they are likewise whirlwinds.

. . . the polar lights radiate in both ultraviolet as well as visible light and are caused by charged particles in the Io torus.

. . . the polar regions are covered by a very high extended layer of vapor.

. . . the extreme upper layer of the atmosphere displays both lightning and flashes of meteors.

Moons and Rings

At least eight volcanos were active on Io in March 1979, hurling material as high as 250 kilometers. The activity of individual volcanos could vary to the point of complete extinction within a period of months.

Deposits of volcanic ejecta pile up so fast that the moon's appearance changes by the month. A large hot spot on Io near the volcano Loki is about 150° Celsius warmer than the surrounding surface.

It seemed at first that there were two fundamentally different types of terrain on Ganymede—one cratered and the other lined by rift valleys. The crust must have been under pressure at some time. With a closer look, however, the landscapes turn out to be mixed together in a complex way.

Callisto is very heavily cratered, one crater practically running up against the next, and has enormous impact basins. Its surface must be several billion years old and basically remained completely unchanged by geological forces. The other moons would look like Callisto were they not subject to change by more recent forces.

Equatorial temperatures on the Galilean moons range between 80 kelvin (−193° Celsius) on the nighttime side to 155 kelvin (−118° Celsius) in the midday sun on Callisto.

Amalthea, a moon with an orbit even nearer to Jupiter than Io's, is shaped like an ellipse of 270 × 170 kilometers. The shape could also be a rhombus. The Voyagers were unable to make any detailed study of Jupiter's many outer moons.

Jupiter is surrounded by a faint ring, the extreme outer edge of which extends 128,000 kilometers from the center of the planet. It consists of a bright narrow segment of lesser density, surrounded by a broad segment covering 5,800 kilometers. Toward the inside is a much thinner layer of material that might extend all the way to the top of Jupiter's clouds.

The ring is best at reflecting sunlight forward, which makes it appear brightest when backlit, and that in turn means that it is made out of particles only a few microns (thousandths of a millimeter) in size. The possibility cannot be dismissed, however, that largish boulders might also be present in between the fine dust particles.

Magnetosphere

Jupiter and the moon Io generate 400,000 volts of power, with an electrical current of 5 million amperes flowing between them inside a magnetic "flux tube."

Sulfur and oxygen in Io's plasma torus cause it to shine brightly in ultraviolet light, but the intensity of its radiation fluctuates widely.

Distributed in various zones inside the complex magnetosphere are hot and cold forms of plasma, made out of protons and oxygen, sulfur, sulfur dioxide, and sodium ions.

Jupiter is a source of radio emission, with a wavelength of several kilometers, which may be caused by oscillations of the plasma in Io's plasma torus.

At a distance of about 12 Jupiter diameters on the side of the planet away from the Sun, Jupiter's magnetic field stretches into a long "magnetotail."

specializing in optical navigation, had an audacious thought. It must be the plume shot into the air by an active volcano! Before long, image analysts were identifying ongoing eruptions in many of the pictures of the last week, coming from a total of eight different craters. Most of the material ejected by the volcanos fell back to the surface, but just enough was escaping to continually replenish the heavy elements in Jupiter's magnetosphere—sulfur, for example.

Io as the most volcanically active body in the Solar System—that was *the* discovery of the mission, front-page news all over the world.

But, unlike all the other surprises of the previous few weeks, the most likely explanation in this case was immediately at hand. It had already been published in the March 2, 1979, edition of *Science* magazine: "Io might currently be the most intensely heated terrestrial-type body in the solar system," three planetary scientists from California had calculated there—and they had predicted that "Voyager images of Io may reveal evidence for a planetary structure and history dramatically different from any previously observed." The reasoning that had now been confirmed so spectacularly by the Voyager photographs was based on the less than favorable position Io occupies between Jupiter and the other large moons. Jupiter's tidal forces pull more powerfully on the near side of the moon than on the side always facing away from the planet. And Io's orbit is constantly being disturbed by the three other major moons. As a result of these combined circumstances, Io is constantly changing shape. The heat generated by the friction of this motion amounts to considerably more than what could have come from leftover radioactive elements inside the planet, which is the driving force of all volcanic activity on Earth. A self-reinforcing melting process leads ultimately to a molten core, which in turn affects the thin surface crust. This explanation finds a place for everything: the bizarre, recent lava formations that immediately efface the craters, hot spots on the surface (picked up by Voyager's infrared sensor), volcanic plumes, and finally both the sulfur and the oxygen released by Io into Jupiter's magnetosphere.

The immediate result of these findings was a special assignment for Voyager 2. The second probe had been launched first, but on a longer route to Jupiter. Now a systematic exploration of Io's volcanos would become part of the basic program, including a 10-hour observation of its volcanic plumes. New closeups unfortunately would not be possible this time, but Voyager 2 would observe Io and the other three moons from a different perspective than Voyager 1, both at a distance and up close. It would also make the first approach to Europa. Meanwhile, in the few months separating the two Voyager arrivals, Jupiter's clouds had already undergone changes. The Great Red Spot was now a monotone reddish orange, more like it was when the Pioneer probes had been there. Everything around the Spot had changed as well, and other swirls, depending on how far they were from Jupiter's equator, had drifted various distances away.

This time, the primary approach to the Galilean moons would come before the encounter with Jupiter itself, and the moon first in line was Callisto. The "other" side was also severely cratered and likewise lacked all sign of geological activity. Changes on Io, in contrast, were already evident just since March. Of four active volcanos then, three—Prometheus, Loki, and Marduk—were still erupting. Pele, the initial volcano discovered by Linda Morabito, had stopped its activity.

The closest approaches to Ganymede, Europa, the small moon Amalthea, and Jupiter all took place on the same day—July 9, 1979. Some thickly cratered parts of Ganymede recall our own Moon or Mercury. Visible in other places are long mountain ranges and trenches that look as if someone had dragged a gigantic rake over the icy surface. Then came the first closeups of Europa, pronounced by a few daring scientists "the most exciting satellite in the Jovian system" in advance of the event. It seemed to be something of a transitional form, between a purely rocky object like Io and balls of ice like Ganymede and Callisto. And it was no disappointment. Looking in many places something like a cracked egg, the surface went on making a more and more powerful impression, with geologists quickly noting similarities with the ice floes on Earth's polar seas. Dividing this bright surface into polygonal plates was an extensive network of dark lines. Nothing of the sort—except perhaps in fantastic nineteenth-century renderings of Mars—had ever been seen in the history of astronomy. And Europa

Voyager 2's view of Callisto from 1.1 million kilometers revealed layer upon layer of cratering, including a large ring-like structure.

possessed another outstanding characteristic. No other body in the Solar System has such a *flat* surface, comparable to a billiard ball.

The Jovian system, in short, was full of records. Here was not only the biggest planet with the strongest magnetic field, but also Callisto, on the surface the oldest and least active body in the Solar System, as well as the darkest (Amalthea), the brightest (Europa), the youngest and most active (Io), and now also the flattest.

All over JPL the mood was festive, and for good reason. Aside from a couple of electronic failures caused by the extreme radiation, both fly-

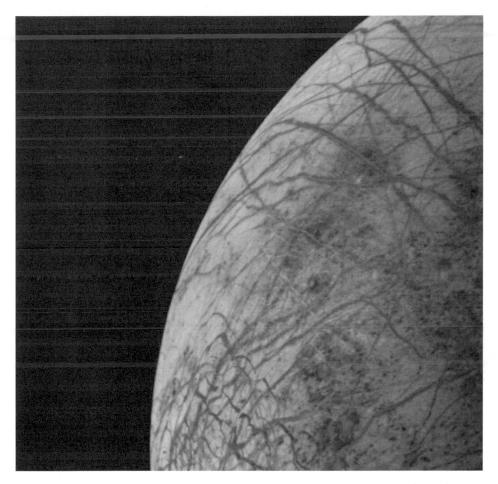

How Voyager 2 saw Europa. Images like this one allowed scientists to learn a great deal about the moon's interior structure. The complete absence of relief structures, for example, spoke in favor of a thick ice crust, and the small number of impact craters indicated a young age. Even the curious pattern of lines was quickly interpreted. The ice crust must have split apart, allowing dark material from inside the moon to flow up to the surface.

bys had gone off practically without a hitch. A new phase in the exploration of the solar system had begun. One NASA official even thought that "a turning point in our cultural, our scientific, our intellectual development" had been achieved. As a farewell token from Jupiter,

Jupiter's southern hemisphere, from the Great Red Spot to the south pole, as seen by Voyager 2. Also visible is a white oval and several other swirling storm centers that influence the large-scale pattern.

Voyager 2 delivered spectacular backlit photographs of several volcano plumes and of Jupiter's narrow ring in a sharply focused version, as well as a modest follow-up on Jupiter's cloud formations. The photographs even contained evidence of a previously unknown Jovian moon, the fourteenth, which would soon be christened Adrastea.

Adrastea ended the immediate excitement of the two Voyager encounters with Jupiter. The probes then traveled on toward Saturn, with Voyager 2 scheduled to venture on to Uranus and Neptune. Preparations had already begun for Galileo, the next mission. Nevertheless, science consists of more than the "instant knowledge" that works so well in the media. JPL experts distill such information out of the data flow in real time, more or less—doing a very good job of it, for scarcely a single important claim had to be retracted later. Still, the real work takes place in the months and years between the spectacular events, and the Voyager mission was no exception. In 1980 a fifteenth moon was discovered on Voyager 1 images. In 1985 another ring was identified. Speculation was lively about the nature of Europa beneath the ice crust. Could it be an enormous ocean of liquid water? Image analysts working on Voyager 2 photographs in 1990 detected significant amounts of lightning on Jupiter's nighttime side. Apparent for the first time years later was how mysteriously systematic Jupiter's atmospheric turbulence is. The storm centers appear exclusively within two narrowly restricted bands, not at all like on Earth. Jupiter had now been visited four times, and still it harbored many secrets.

The time had come, however, for the kind of systematic investigation of the planets which could be undertaken only from a position in orbit around them. Launched in 1989, Galileo would take six years to reach Jupiter. But it was not only the journey itself that was long. From the first plans being made around 1974 to liftoff, 15 years had passed— twice as long as originally hoped. The Galileo odyssey began on Earth.

What Became of the Pioneers and Voyagers?

They were pioneers in the exploration of our Solar System, revolutionizing our thinking about Jupiter and the other large planets—but that was not the end of the long journey for the Pioneer and Voyager probes. They are the first (and so far the only) emissaries from Earth on trajectories that will take them out of the Solar System—a circumstance that scientists exploited as best they could. Contact could be maintained with the Pioneer probes until the mid-1990s, with Pioneer 10, in particular, proving to be nearly indestructible. As late as mid-2000, its radio signal could still be picked up in the sky as a point source of radio emission. By late 2000, however, no more reliable contact could be established.

The future belongs to the Voyagers, which are on faster trajectories. On February 17, 1998, Voyager gained the distinction of having traveled farther than any other celestial object ever launched from Earth. And both Voyagers should be able to go on working until 2020. After the Voyager 2 Neptune flyby, the project was renamed the Voyager Interstellar Mission (VIM), expressing the scientists' hopes that the probes could continue transmitting data after they had escaped the Sun's sphere of influence altogether and were headed into interstellar space.

The "edge of the Solar System" is no mere mathematical construction. It indicates the area that is dominated no longer by the solar wind, but instead by the sparsely distributed gas of the Milky Way. Theorists have developed their understanding of this boundary in detail, and observers have provided schematic proof of its existence, but nothing can beat actual observations *on the scene*. At the end of 1997, 20 years after having been launched, both Voyagers clearly remained in the solar wind, as evidenced by the supersonic particles streaming from the Sun. Voyager 1 had already traveled a distance of 67 astronomical units (AUs), or 10 billion kilometers, from the Sun, extending the distance each year by 3.5 AUs, or 500 million kilometers. Voyager 2 had traveled 8 billion kilometers. At some point, particle detectors on the Voyagers will register a decline in the speed of the particles and other characteristic changes as the solar wind is slowed to subsonic speeds by the resistance of the interstellar gas it confronts there. This will be the "termination shock," theorized to be located about 85 AUs from the Sun, so that Voyager 1 will probably pass it in 2002 or 2003. By accident, both Voyagers are headed in a direction in which the boundary of the heliosphere (the Sun's sphere of influence) may be particularly close. Because the Sun is in motion relative to interstellar space, the heliosphere is reduced in this direction, while in the opposite direction it may extend into a sort of tail.

Beyond the termination shock lies the so-called *heliosheath,* in which space continues to be dominated by solar particles, although they are moving at a much reduced speed. This zone could stretch for dozens of AUs, and getting through it could take several years. At that point, however, the *heliopause* (the outer edge of the heliosheath) will have been surpassed and the heliosphere left behind for interstellar space. Scientists estimate the final boundary to be between 110 and 160 AUs from the Sun. When it is crossed, the Voyagers will have achieved their ultimate goal, inaugurating a new phase in the history of space flight.

There will be no pictures of the event, unfortunately. The Voyagers' cameras have been shut down for years now, with what remains of the steadily dwindling energy supply from the old radioisotope batteries distributed among five selected field and particle experiments. Magnetic fields and plasma waves, plasma and low-energy charged particles, and cosmic rays are under systematic observation. It seems that the Voyagers already made one critical discovery in the early 1990s. The Plasma Wave System onboard both spacecraft began picking up intense, very low frequency radio waves, which are obviously produced at the heliopause. The most powerful radio source in the solar system, it cannot be detected at all from Earth.

Most Voyager data travel to Earth in real time, at 160 bits per second. Once a week, however, data from the Plasma Wave System are put on tape, to be transmitted as the occasions present themselves and received by a 70-meter antenna. Yet, nothing lasts forever, and in the years 1998 and 2000, respectively, the scan platforms on Voyager 1 and 2 were to be shut down, ending the use of the UV spectrometer for astronomical readings. Attitude control will quit functioning around 2010, requiring more and more of the instruments still active at that time to be finally turned off. In 2020 neither Voyager will have enough power left to run even a single scientific instrument. That will mark the end of the scientific mission, but even then both travelers will continue on their way . . . .

Will there be probes in the future that explore beyond our Solar System? NASA is at least considering the possibility. Probes made especially for interstellar space could make it to the neighborhood of a nearby star in a manageable amount of time. What they call for is a completely new propulsion system. In January 1998, NASA head Dan Goldin challenged the astronomers and engineers of the world to devote some thought to what might become a reality by the end of the twenty-first century.

From an Idea to a Billion-Dollar Spacecraft

The beginnings of the Galileo project go back to the era of the Pioneer probes—the Voyager project had not even started. In 1974, however, planetary scientists already knew what they wanted for the third generation of long-range probes. A study by the Space Science Board of the National Academy of Sciences took up the question of NASA's priorities in the decade from 1975 to 1985. "We recommend that a significant effort in the NASA planetary program over the next decade be devoted toward the outer solar system. Jupiter is the primary object of outer solar system exploration." The chemical composition of its atmosphere and its physical state should be the focus of research in particular, as well as the composition and state of the moons and the magnetic field and charged particles. To do that, it would be necessary to have both an orbiter and an atmospheric probe. NASA already had a pair of studies at the time, based on existing probes. Scientists at the Ames Research Center were considering the possibility of using a modified Pioneer-type probe to descend into the atmosphere and, at the same time, maintain an orbiting spacecraft to conduct a systematic investigation of Jupiter's magnetosphere. Another working group favored a probe more along the lines of the Voyagers, with attention focused primarily on the moons.

NASA combined the two concepts in 1976, proposing a kind of super-Voyager spacecraft that could release a probe to explore Jupiter's atmosphere. The mission was now called JOP, for "Jupiter Orbiter with Probe." The Jet Propulsion Lab would be responsible for the probe, while Ames Research Center would produce the orbiter. Twenty years later, this division of labor would become politically charged in a curious way. JOP had not yet been approved, and in the months leading up to the Voyager launch, the action shifted to Washington. NASA managers went through endless congressional hearings, and several votes had to be taken before the project was passed. The formal "new start" of the Galileo mission, as it was quickly but fortunately renamed, was set for July 1, 1977, with the selection of scientific instruments to pack on board to be decided in August. The timing was ideal. Many members of the Voyager team, with their probe now on the launch pad, could continue right on with Galileo. Experience gained building the Voyager probes and their subsystems could be carried over. The same was true at Ames, where technicians were putting the finishing touches on capsules to be launched with the Pioneer probes in 1978 and designed to descend gently down into the Venusian atmosphere. What they had learned could be applied to the Jupiter probe.

The great progress of Galileo over the Voyagers lay, first off, in the advances made in the meantime by both NASA-based and university scientists. After all, seven years had passed. And there was no doubt that an atmospheric probe could deliver completely new insights about Jupiter's atmosphere. For the first time, it would be possible to analyze it directly. Data on temperature and barometric pressure, as well as chemical analyses, would be much more precise because of the most recent advances in the development of remote sensors. Much of the instrumentation on Galileo was adapted from Voyager, but in each case there were distinct improvements. Some of the devices, such as the camera, looked the same as their forerunners. Galileo's camera would use the same optical system as the earlier probes, but this time it would have a semiconductor CCD chip instead of television tubes. Steady improvements can also be traced in the particle detectors available from the Pioneers to the Voyagers to Galileo. The primary focus had at first been particle energies, but now it would be possible to quantify the chemical elements involved in particle motion as well as determine in detail how they moved.

Not only technology, but the conception of the mission altogether promised major improvements in data quality. The four previous probes had raced through the Jovian system on a single course, taking snapshots and sampling the magnetosphere. Galileo, in its orbit around the planet, would pass through the same space over and over. This would make it possible for the first time to distinguish temporal and spatial changes, a fundamental problem in the study of all magnetospheres. It was even supposed to be possible to make an excursion far into the tail of the magnetosphere, just over one hundred Jupiter diameters away from the planet itself. Worlds would open up on each of the Jovian moons. They would be visited repeatedly, making much closer flybys than the Voyager probes. In addition to the new camera, Galileo had a complicated instrument called NIMS (Near Infrared Mapping Spectrometer) that could register not just spectra, but also infrared images in a number of wavelengths.

The plan was for Galileo to lift off with a space shuttle in January 1982, carried aloft by a rocket with a specially modified main stage. The trip to Jupiter would be relatively direct, with only a quick visit to Mars to pick up acceleration. Scheduled arrival at Jupiter: the mid-1980s.

But nothing would go as planned. Manned spaceflight was one major factor. It was NASA's unswerving resolution that all aspects of space transport, including launches of planetary probes, would be taken care of by shuttles. Toward the end of the 1970s, however, the shuttle program was in trouble, struggling in particular with the complicated main propulsion system. The first launch would not come until April 1981, and there would be no room for Galileo for years, at least until 1985. Something else had happened, too. The White House had changed hands, resulting a whole new set of priorities for the space program. Exploring the planets meant nothing to the Reagan administration, and Galileo was among the programs slated for termination. The "new start"of 1977 was meaningless now, because government programs are subject to annual approval, no matter how much money has already been spent.

Nr. 189 / Donnerstag, 18. August 1983   **WELT UND WISSEN**

Galileo darf nicht sterben

Seit vielen Jahren in Vorbereitung befindliches Weltraumprojekt durch rigorose Sparpläne der Reagan Administration gefährdet / Öffentlichkeitskampagne gestartet

VON UNSEREM MITARBEITER WOLFGANG ENGELHARDT

Die neue Wirtschaftspolitik des amerikanischen Präsidenten mit ihren drastischen Einsparungen in praktisch allen Lebensbereichen macht auch vor dem Etat der Raumfahrtbehörde NASA nicht halt. Besonders betroffen von den Sparplänen der US-Administration ist das Galileo-Projekt der NASA, das die Entsendung eines Orbiters und einer Atmosphärensonde zu dem Riesenplaneten Jupiter vorsieht. Nach den fantastisch erfolgreichen, im rasenden Vorbeiflug aber nur sehr kurzzeitig beobachtenden Voyager-Sonden könnte der Jupiter-Orbiter Galileo den orangeroten Planeten mit seinen vier großen Monden einige Jahre lang erkunden. Mit der Atmosphären-Sonde könnte die überdimensional große und turbulente Gashülle des Riesenplaneten erstmals direkt „vor Ort" untersucht werden.

Das Galileo-Projekt kämpft eigentlich schon seit seiner Begründung in der Mitte der 70er Jahre um das Überleben und seine endgültige Konzeption. Mehrere Budget-Kürzungen in der Vergangenheit und vor allem Schwierigkeiten mit dem als Startfahrzeug vorgesehenen Raumtransporter verschoben das Startdatum von 1981 bis schließlich 1985. Wenn dieser letzte Termin nicht eingehalten werden kann, dann wird Galileo überhaupt nicht mehr starten, weil dann die Bahnverhältnisse von Jupiter und Erde zueinander immer ungünstiger werden für eine solche Mission.

Angesichts dieser dramatischen Zuspitzung der finanziellen Situation um Galileo hat sich in den USA eine Art Bürgerinitiative gebildet, die unter Leitung der bekanntesten Astronomen und Raumfahrtexperten eine Öffentlichkeits-Kampagne gestartet haben, um den Jupiter-Orbiter doch noch zu retten. Kern dieser Aktivitäten ist eine Briefaktion, bei der

neller und auch wissenschaftlicher Hinsicht gleichkäme. So will man z. B. auch das Zentrum für bemannte Raumfahrt in Houston (Texas) schließen. Alle Aktivitäten könnten – so meinen die amerikanischen Spar-Kommissare – am Cape Kennedy konzentriert werden, wo der Raumtransporter startet und demnächst auch landet. Der Shuttle ist überhaupt das einzige große Raumfahrt-Projekt der USA,

das von den Sparplänen ausgenommen wurde, was sicherlich auch auf militärischen Nutzanwendungen zurückzuführen ist, die nach den Plänen der US-Luftwaffe damit realisiert werden sollen. In dieses Bild paßt die Tatsache, daß die amerikanische Luftwaffe erstmals me\ır Geld für Raumfahrt-Unternehmen erhält, als die zivile Astronautik-Behörde NASA. Die Air Force baut damit vor allem eine neue Startanlage für den Raumtransporter in Kalifornien, von wo aus die sogenannten polaren Umlaufbahnen erreicht werden können. Aus diesem Orbit heraus kann ein Aufklärungs-Satellit in regelmäßigen Abständen alle Punkte der Erdoberfläche einsehen.

Mit dem völligen Abbau aller Programme zur Planetenforschung gehen auch Pläne einher, das dafür spezialisierte Forschungszentrum der NASA in Pasadena bei Los Angeles zu schließen, bzw. zu einer Versuchsstation der US-Luftwaffe umzugestalten. Einige hundert hochspezialisierte Astrophysiker und Raumfahrt-Ingenieure des „Jet Propulsion Laboratory" würden arbeitslos, ein erprobtes, eingespieltes Experten-Team auseinandergerissen. Ein „Nebenprodukt" dieser radikalen Sparpläne der US-Regierung wäre auch die Stillegung der drei Antennenstationen des „Deep Space Network" in Kalifornien, Spanien und Australien, die bislang die Funkverbindung mit den amerikanischen Planetensonden gewährleisten. Damit wäre dann natürlich auch der weitere Funkkontakt zu der Raumkapsel Voyager-2 abgebrochen, die im Jahr 1986 noch an dem Planeten Uranus und 1989 am Neptun vorbeifliegen soll.

GALILEO
FLUGBAHN ZUM JUPITER

50 TAGE — MARS
VENUS — ERDE
SONNE
1 AU

"Galileo Must Not Die": German newspaper headline. In the 1980s, voices from all over spoke out against canceling the Galileo project, which was threatened by U.S. budget cuts.

Galileo was 90 percent ready! And it turned out not to be a time when scientists would stand by while important programs were canceled. Led by some of the most prominent astronomers and experts on space travel, a kind of popular initiative to save Galileo got under way. The White House received hundreds of thousands of letters from citizens, including one from James van Allen, the discoverer of the radiation belt around Earth and member of the first U.S. satellite project, Explorer 1.

Canceling the Galileo project would have affected European scientists and technicians as well, especially in Germany. NASA had invited the Federal Republic of Germany to help work on the drive module for the attitude control, course correction, and retro-rockets, including all of the thrusters and ancillary systems. This was the first time NASA had given the contract for one of the central systems of a space probe to any foreign firm. Messerschmitt-Bölkow-Blohm (MBB), today part of Daimler Benz Aerospace (DASA), had been chosen as partner. MBB had made a name for itself developing the attitude control system for *Symphonie,* the German-French communications satellite. Moreover, three of Galileo's 17 scientific instruments, including devices for both the orbiter and the probe, were to be developed in Germany. German scientists were also involved in the development of many of the other instruments and would later take part in the data analysis. Yet, in this kind of bilateral collaboration, no money flows across the Atlantic. At DM 40 million, the German share of the project accounted for ten percent of overall costs, paid by the Federal Republic of Germany in exchange for a chance to fly along, so to speak.

In short, there was good reason in Germany to follow developments in the U.S. with concern: "Galileo Must Not Die," ran newspaper headlines. And, with the astronomer and media star Carl Sagan reportedly playing a central role, the mission was indeed saved in 1983. Shuttle liftoff was now planned for 1986. A powerful Centaur rocket main stage would hurl Galileo from Earth orbit directly to Jupiter, without detours to other planets. Scheduled arrival: August 1988. It looked as if everything was set to go. A model of the atmospheric probe was tested from an altitude of 30 kilometers, and it landed as planned nine minutes later in the desert at White Sands Missile Range in New Mexico. Conditions corresponded approximately to the probe's descent toward the gas giant, but they did not reproduce the drastic braking that would ensue upon the probe's initial contact with Jupiter's outer atmosphere. That phase of the mission was impossible to simulate in any laboratory on Earth. The mission picked up a special bonus in 1984, when celestial mechanics astronomers determined that Amphitrite, a particularly large asteroid (number 29, officially), was going to approach Galileo's route through the Asteroid Belt. To steer toward the object for a closer look would cost some fuel and postpone arrival until December 1988, but the chance to capture the first photographs ever of a "minor planet" made it worth the sacrifice. At 200 kilometers in diameter, Amphitrite represented an imposing object, the twelfth brightest asteroid in Earth's night sky.

With the launch date May 20, 1986, approaching it seemed that nothing more would stand in the way of the Galileo project. Ten years after the project's inception, the journey to Jupiter was about to start. The outer Solar System was more enticing to astronomers than ever, with both Voyagers having made major discoveries in 1980 and 1981 about Saturn and its rings and moons, which easily rivaled the excitement over Jupiter. Still fully functional in January 1986, Voyager 2 arrived at distant Uranus, and that planet, like the others, was revealed to be a strange world surrounded by complicated rings and bizarre moons.

Then, with the Uranus flyby still in progress, came January 28, 1986, the blackest day in NASA's history. After a series of mistakes and misjudgments, and despite unusually chill nighttime temperatures at the Kennedy Space Center, the okay was given for liftoff of the space shuttle Challenger, with seven astronauts on board and a data relay satellite as payload. The twenty-fifth launch of the shuttle, or in the official numbering system STS 51-L, ended in disaster. Hot gas that had escaped from one of the solid-fuel rockets destroyed the shuttle's main fuel tank, and it exploded. The Challenger was torn apart by aerodynamic forces. There was no way to save the astronauts. And for two-and-a-half years, the space program came to a halt.

But there was even more. Suddenly, in June of 1986, the Centaur main stage needed for the launch to Jupiter became unavailable! This rocket stage, fueled by liquid hydrogen and oxygen, had proved itself on unmanned flights, for example in the Voyager launches. Now, despite $700 million in development costs already spent, NASA had resolved that such an explosive mixture would no longer be tolerated on shuttle flights, should it be decided that they would fly again. A Centaur would represent an unconscionable risk in case of an aborted launch or emergency landing. American planetary scientists were devastated. Now there was simply no way to get the probe, finished and ready to go, to its destination. There were no longer any transport systems for Galileo and other planetary probes. Experts considered combining several "permitted" stages, but it quickly became evident that, with Galileo added in, they would be too heavy.

The ultimate solution first began to appear in September 1986. It used only existing technology—plus the forces of nature. It was probably the only way Galileo would ever get from Earth to Jupiter, and it was christened VEEGA, for Venus-Earth-Earth Gravity Assist. The trick involved launching Galileo toward Venus, rather than Jupiter. The only thing needed for that was an inertial upper stage, which was available, using a solid-fuel mixture that had been approved for the shuttle. From one Venus and two Earth flybys, Galileo would gather sufficient force to make it all the way to Jupiter. It was the same method that had been used to "slingshot" the Voyagers from planet to planet, saving ample quantities of both fuel and time. The spacecraft can actually rob a planet of an infinitesimal portion of its momentum, which makes a planetary flyby the equivalent of an extra rocket stage. A good one. Had Galileo burned all of its 932 kilograms of fuel at once, it would have added 1.6 kilometers per second to its speed. The Venus and Earth flybys, taken together, would cause its speed to increase by six times that amount—for free. The downside was that the detour through the inner Solar System would significantly lengthen Galileo's trip to Jupiter. Assuming liftoff in the fall of 1989, according to JPL calculations, arrival could scarcely be imagined before 1996. No one knew yet when the shuttles would fly again, but certainly not before 1988. Galileo would have to try for a place on one of the first flights.

The second problem with VEEGA involved going from Earth to Venus. Galileo had not been designed to come so close to the Sun, so the new route called for some rebuilding and the installation of special heat shields. Anything was better than just depositing the spacecraft in the National Air and Space Museum, however, and the new plan was approved before the end of 1986. "The phoenix has risen from the ashes," in the words of one trade journalist congratulating NASA for its innovative spirit. Not only had Galileo been saved, a whole list of sites to explore had been added to its itinerary. Such a combination of remote sensing instruments had never been sent to Venus, and it would even be possible to arrange a couple of unusual investigations during the two close encounters with Earth and the Moon. The rendezvous with Amphitrite was lost, but in exchange Galileo would fly through the Asteroid Belt twice instead of once—and both times, at some cost of fuel, it would be possible to steer over for an encounter with the small asteroids.

Adapting Galileo to the new mission made a lot more work for project engineers. In early 1987, the probe was sent back to California from the Kennedy Space Center in Cape Canaveral, Florida—on a truck convoy. Most of the scientific instruments had already been removed from the probe and sent back to the various institutions in the U.S. and Germany where they had been built, for either storage or improvement. The scientists had not exactly expected to see their creations again. And some of them were frustrated for reasons that went beyond the new delays. The instruments, which had been ordered a decade ago now, often had been built by people who were no longer working together. Galileo itself arrived back at JPL on February 21, 1987, where it was completely disassembled. VEEGA called for an enormous number of technical details to be altered. In the summer of 1987, before Galileo was ready to be put back together, it presented a truly sorry sight in JPL's clean room. And it still was not clear whether

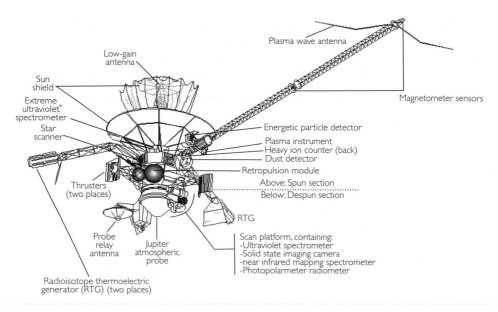

Low-gain
antenna

Sun
shield

Extreme
ultraviolet
spectrometer

Star
scanner

Plasma wave antenna

Magnetometer sensors

Energetic particle detector

Plasma instrument
Heavy ion counter (back)
Dust detector

Retropulsion module

Above: Spun section
Below: Despun section

Thrusters
(two places)

RTG

Probe
relay
antenna

Jupiter
atmospheric
probe

Scan platform, containing:
-Ultraviolet spectrometer
-Solid state imaging camera
-near infrared mapping spectrometer
-Photopolarmeter radiometer

Radioisotope thermoelectric
generator (RTG) (two places)

Sketch of the Galileo spacecraft.

it would win the stiff competition for tickets on the fall 1989 shuttle. One rival was Ulysses, the solar probe built by the European Space Agency (ESA), which NASA had agreed to launch and which would also be conducting U.S.-sponsored experiments. Ulysses was supposed to lift off as soon as possible.

Officials from NASA and ESA got together in April of 1987 to decide on the launch order. Galileo would take the early date, as planned, with Ulysses following a year later. Ulysses would take a faster route over the Sun's polar region, making it possible to expect the first data to arrive in 1994. Had Ulysses and Galileo traded places, Galileo's mission to Jupiter would have gotten under way not in 1996, but 1998! The quarrel over the best seat picked up again a few months later, when it seemed that the resumption of shuttle flights would be delayed. In December, the decision was final: Galileo would lift off for Jupiter in October 1989. On the way, it would pass by not only Venus, Earth, and the Moon, but also the asteroids Gaspra (officially, number 951) and Ida

The original Galileo space probe inside a large testing chamber.

(number 243), before finally arriving at its destination on December 7, 1995—about 20 years after the JOP project first got under way. The scientific program would be carried over almost unchanged. Of course, what could not be maintained were the costs. Just adding the new systems that had become necessary for the probe raised its cost from $700 to $900 million. The mission had lasted much longer than anticipated, sending those costs up from $200 to $500 million. Still, a somewhat modernized Galileo would be sent off now because of improvements made to the instruments before it was reassembled in June 1989.

Once again, it seemed that all was set for liftoff. Then, at the end of 1988, German engineers made a shocking discovery on a completely different space project, again threatening the entire Galileo mission. Everyone knew that MBB had gotten the contract for Galileo's primary and secondary thrusters because MBB jets had worked so well on the *Symphony* satellites. The Bavarian company has also been responsible for the propulsion systems on later Earth-orbit satellites; for example, the German TV-SAT 1. This direct-broadcast television satellite, which was supposed to ring in the age of satellite TV in Germany, had become a joke in 1987 when the solar-cell paddles would not open. Someone had forgotten to remove the clamps before liftoff. The expensive satellite had enough power for only a couple of experiments—including running tests on MBB's attitude control jets, closer relatives of the small thrusters on Galileo. It turned out that they have a tendency to overheat, so that they could destroy themselves! "We would certainly have encountered problems," Galileo's project manager, R. Spehalski, was forced to concede, "had we made the May 1986 launch date."

Aside from the main engine, delivering the 400 newtons of thrust needed for braking the space probe into orbit around Jupiter, which was operating flawlessly, Galileo also had these 12 attitude control thrusters. Not only did they have a tendency to overheat, while accounting for only 10 newtons of thrust, they were expected to use up a

Final assembly of the Galileo spacecraft.

full third of the fuel supply over the course of the mission. And now tests being carried out at the rocket test center in Lampoldshausen, Germany, were showing signs of a "hot start" problem that might destroy the thrusters within a few seconds of ignition. The defect was traced to an excessive build-up of helium bubbles in the fuel lines, and that threatened to diminish thrust by five percent. Given the precise calculation of fuel supply, that would have meant giving up a number of maneuvers, for example, the approach to the asteroids or course alterations inside the Jovian system.

JPL took delivery of the modified jets on April 4, 1989. The overheating problem had been solved, but full thrust was still not available. Instead of letting them burn seven to eight minutes, as the mission plan frequently called for, they could be fired for only a few seconds at a time. Maneuvers that had been estimated in hours now stretched over days. Fuel consumption would certainly increase; by how much, no one knew. On the other hand, ample reserves had always been part of the plan. On May 16, following another cross-country trip by truck, Galileo arrived once again at Kennedy Space Center. The orbiter was first tested on its own, and then engineers reinstalled the atmospheric probe. The inertial upper stage was finally put in place in July, and in August Galileo was transported to the building where the payload was being readied for launch with the Space Shuttle Atlantis.

Again during these few weeks, planetary scientists found themselves under the spell of distant worlds. Voyager 2 had reached Neptune, the fourth destination on its itinerary, almost completing the grand tour on its own; only a jaunt to Plato had not proved possible. Pictures of Neptune, the geysers on its large moon Triton spewing dark trails of smoke into a thin atmosphere, made more publicity for Galileo. Liftoff in mid-October now seemed assured—the shuttles had been flying reliably since September 1988. Nothing short of a natural disaster could interfere with the launch—or a court of law. . . .

Chapter 2

Discoveries Under Way

The Trip to Venus

September 1989: Galileo stands in the payload hold of the Atlantis space shuttle on the launch ramp at Kennedy Space Center in Florida. The launch window for the planned VEEGA route to Jupiter runs from October 12 to November 21, but the later the launch after October 12, the more fuel will be required to approach the two asteroids Gaspra and Ida. Shuttles have now flown five times since the tragedy, and operations are flawless. And it is not the first time for an interplanetary probe to be launched by a shuttle. Magellan, a radar orbiter dispatched to Venus in May, had been lifted into space on a shuttle. But this is the first time a shuttle will lift off carrying radioactive material.

There were about 11 kilograms of plutonium in the radioisotope batteries designed to power Galileo. In fact, the situation was not different than it was with both the Pioneers and the Voyagers, which were launched (on one-way rockets) between 1972 and 1977 without attracting particular attention. But times had changed. In the decade since, two major nuclear accidents had taken place in the U.S. and the Soviet Union, and Challenger had gone down. The mood was different now, and Atlantis, with its purportedly dangerous cargo—the "plutonium probe"—on board looked destined to set an example. Several American organizations filed a complaint in court, seeking an injunction against the launch. After 15 years of continuous back and forth, Galileo was finally cleared for liftoff. The launch window that would accommodate the complicated VEEGA trajectory was coming ever closer. Now, this.

Explicit White House approval is required anytime a probe carrying radioisotope batteries is launched. This had been taken care of in mid-September. How was it that the political officials involved, all of the scientists working on the project, pro-space-travel groups, and even a few organizations that are usually critical of the government had found no cause for concern in the launching of 11 kilograms of plutonium-238-dioxide into space on a shuttle? First, radioisotope batteries—more precisely, "radioisotope thermoelectric generators," or RTGs—are indispensable if a large space probe is to be supplied with energy as far from the Sun as Jupiter. Second, this is a tested technology, under the control, incidentally, not of private research interests, but of the U.S. Department of Energy. By 1989, the number of American space flights involving RTGs was pushing two dozen, and the devices themselves had been constantly improved over 20 years. RTGs had never been responsible for an accident, although two launches involving RTGs had failed for other reasons—and both times the "casing" had done its job. No radioactivity was released. Apollo 13's RTGs sank into the ocean. In 1968, the battery pack for a Nimbus weather satellite was even recovered and reused.

The most important thing to know about what RTGs are really like—and they have nothing in common with the indeed dangerous nuclear reactors the Soviet Union used to send up in military satellites—is that the short-lived plutonium isotope plutonium-238 is not weapons grade and cannot explode. Plutonium decays naturally, producing heat. The heat is then either used with a thermal element to generate electricity or put to other uses requiring heat. Since the plutonium exists as dioxide, it forms an insoluble ceramic that can be broken apart only under severe impact. And even then, it breaks into large blocks, *making no dust*. This last characteristic is the most important. A source of alpha radiation, plutonium-238 causes cancer in humans only if it is inhaled and allowed to lodge in the lungs. Ordinary skin is protected from the radiation by a layer of dead cells. The plutonium dioxide in every RTG consists of 18 separate modules of four "pellets" each, with each individual module protected individually against impact, and each with its own heat shield. Even in one extremely severe accident, only a few modules were damaged. The radioactive material is enclosed inside many layers of extremely heat- and impact-resistant materials (including iridium and graphite).

Comprehensive tests involving spectacular crash experiments have shown that these protective steps are adequate for virtually every accident scenario. An explosion with a force ten times that which destroyed the Challenger would leave the RTGs unscathed. In theory, at only two points in the Galileo mission could an accident occur in which radioactivity would be released and reach Earth: at takeoff and on the two swings by Earth in 1990 and 1992. The VEEGA route called for the spacecraft to re-approach Earth to pick up acceleration.

An explosion on the launch pad or during the first few minutes of flight represented a relatively real possibility—thus the intensive precautionary measures involving the special packaging of the plutonium. The chance of Galileo crashing full speed into Earth on one of the swingbys, in contrast, was extremely remote.

Until just before dipping in to pick up the assist, the spacecraft's trajectory called for it to remain thousands of kilometers from Earth. Even during the period when radio contact would be interrupted and it would be impossible to transmit commands, nothing could happen to Earth. Flight planners spent precious fuel in order to keep the probe as far away from Earth as possible until just before the swingby, when the optimal course would be selected. Additional care was taken so that at no point during the entire mission would Galileo be aimed on a course toward Earth. That flight controllers would have no trouble orchestrating such maneuvers had been amply proved by their impressive performance guiding the Voyagers from planet to planet.

Only in the extremely unlikely event of a swingby accident, in short, was there even a chance that plutonium would escape from the RTG casing into the atmosphere. Even then, according to comprehensive studies, the increase in radioactivity against the natural background would be too slight to measure. There remained those, nevertheless, who either could not or would not see the logic of the argument. Galileo even showed up on the agenda of the German parliament, prompting the national government to declare that it bore no responsibility for the project as concerned safety matters. Ultimately, the renowned astronomer and science writer Carl Sagan took up his pen. As a trenchant critic of nuclear arms, he was not one to be suspected of speaking for the "plutonium lobby." Sagan vehemently castigated the uninformed critics, arguing in support of Galileo. But the court had yet to rule. It was October 10, a day after the Atlantis countdown had begun, before the good news finally arrived. Judge Oliver Gasch had rejected the complaint. He saw no reason to believe that

NASA had understated the risks. An injunction against the launch, he ruled, would do the United States more harm than good.

The Mission Begins

Final clearance for the thirty-first launch of the shuttle, in this case Atlantis, flight number STS-34, unfortunately did not mean that Galileo's long journey could at last get under way. Almost simultaneously, NASA announced that the computer responsible for operating the shuttle's main engines was not functioning. The Main Engine Controller in the shuttle's tail had to be replaced—but then things would finally get serious. October 18, 1989, was an historical day in several regards. It was the day the East German head of state Erich Honecker stepped down, the day San Francisco was struck by a major earthquake—and the day Galileo took off. A control center had been damaged in the earthquake, nearly delaying the launch again, but soon the countdown was resumed and, during a lull between two storm systems, the big moment finally came. Only three minutes had passed since the daily launch window had opened. "We have ignition—and liftoff of Atlantis and the Galileo spacecraft bound for Jupiter." Making the announcement at 11:53 A.M. local time, the NASA spokesperson was by no means the only one celebrating.

The launch itself was made in an unusual direction, with a mandatory trajectory angle of exactly 34.4°. The two-part inertial upper stage sent along to power Galileo was so weak that it would have to depart Earth orbit 295 kilometers above the ground exactly in the plane Venus would occupy at the time. The time came 6 hours and 21 minutes into the flight. Scarcely recognizable mounted on the large rocket stage, Galileo was gently catapulted out of the shuttle payload bay. Weighing precisely 20 tons, the probe and its rocket drifted slowly away from Atlantis. An hour later and a safe 80 kilometers away, the first stage burned

The Scientific Goals of the Galileo Mission
(defined prior to launch)

Exploring the Jovian atmosphere
  analyzing its chemical composition
  exploring its structure to a depth of at least 10 bars
  determining the nature of the cloud particles
  determining the radiation flux
  investigating atmospheric currents and their dynamics
  investigating the upper atmosphere and ionosphere

Exploring the Jovian moons
  characterizing their morphology, geology, and surface physics
  researching their respective surface mineralogy
  determining gravitational and magnetic fields
  investigating their atmospheres, ionospheres, and extended gas
  clouds
  investigating the interaction between Jupiter's magnetosphere and the
  moons

Exploring the magnetosphere (to a distance of 75 Jupiter diameters)
  characterizing the energy spectrum, composition, and dynamic patterns of energetic particles throughout the magnetosphere
  characterizing the magnetic field
  characterizing the plasma, including plasma waves

Five Specialties of the Galileo Mission

- The atmosphere of one of the outer planets would be explored directly for the first time.
- For the first time an orbiter would circle one of the outer planets.
- The moons would be explored from a perspective as much as 100 times nearer than ever before.
- The Jovian environment would be systematically examined for at least 2 years.
- The instruments are all capable of much more than their predecessors.

Costs

Total costs, 1977–1997, for the U.S.: $1.354 billion ($982 million for development, construction, and the first 30 days in space plus $462 million for operations from 1989 to 1997). Further U.S. costs incurred for the launch and the use of the Deep Space Network for communications. Additional costs to foreign partners (especially Germany): $110 million—recognized by NASA as "very substantial."

for a total of 150 seconds, then the second stage for 105. Galileo was now moving much more slowly relative to the Sun and the Earth, causing it to "fall" toward Venus in the inner Solar System. The second stage would be needed for one additional maneuver. The probe had to be oriented on the correct angle to the Sun, activating its on board solar sensor, and it had to be left spinning at three revolutions per minute. Forty minutes later, the second stage was cast off, and Galileo was alone.

Not only in space was Galileo alone. Communications with Earth would have to be reduced to an absolute minimum. For reasons having to do with its heat shield, the probe had to maintain a very precise orientation toward the Sun, and the high-gain antenna had to be kept closed. With the low-gain antenna pointed away from the Earth, it was possible to maintain contact at all only because of a less powerful auxiliary antenna installed specifically because of the coming trip to Venus. None of this, however, was going to prevent scientists from making the most of Galileo's Venus flyby. A magnetic tape data recorder with a capacity of 900 kilobytes made it possible to take the equivalent of 200 pictures, with the data to be picked up on the first Earth flyby. The Venus trip would serve not merely to pick up a gravity assist and test the orbiter's instruments on a real planet, but would also contribute to scientific knowledge. And nothing else was of greater interest to researchers at this point. The Galileo project had been under way for 15 years since it was first conceived. It had been seven years since the originally scheduled launch date.

Galileo carried a payload of 148 kilograms. Of that total, 118 kilograms belonged to the ten experiments to be conducted from the orbiter (which itself weighed 2,380 kilograms at launch, including 1,089 kilograms of fuel). The atmospheric probe's six experiments accounted for 30 kilograms of payload (with the probe adding 338 kilograms to overall weight). The Galileo orbiter combined the advantages of the spin-stabilized Pioneer spacecraft, which rotated constantly around its

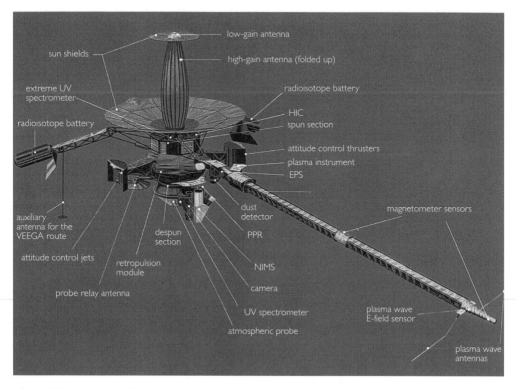

The Galileo spacecraft as it was prior to the first Earth flyby, before any attempt had been made to open the high-gain antenna.

own axis, and the Voyager, which went through a series of spatial orientations, each centered on one of three axes. The Pioneer's constant rotation made it ideally suited to get a "panoramic view" of the particles and fields it passed through, while the ingenious camera platforms on the Voyager's remote sensing instruments were able to hold a stable focus on distant objects. The upper part of Galileo, the "spun section," rotated typically at 3 rpms, while the 11-meter-long magnometer arm swung regularly through space. The engines, most onboard electronics, the power supply, and the high-gain antenna all rotated along. The spun section carried the plasma wave system (PWS), an instrument for studying plasma waves, a magnetometer (MAG), an energetic particle detector (EPD), a plasma detector (PLS), and the dust detector (DDS).

The Galileo Orbiter

Mass:	2,223 kilograms at liftoff, including 118 kilograms of scientific payload and 925 kilograms of useable fuel
Stabilization:	spin-stabilized, normally at 3 rpms (10 rpms when the main engine is in use to force the fuel against the sides of the tank)

Propulsion System:

Thrusters:	one main engine (400 newtons of thrust) 12 small thrusters (10 newtons each) on two extension arms (2 meters long)
Fuel:	364 kg monomethyl-hydrazine (MMH)
Oxidizer:	595 kg nitrogen tetroxide, enriched with one percent nitric oxide (MON-1)
Pressurizer:	2.8 kg helium
Tanks:	two, each with a diameter of 74.8 cm

Estimated fuel consumption:

Small thrusters:	321 kg for repeated deployments from 1989–1998
Main engine:	40 kg for the Orbit Deflection Maneuver, 1995 377 kg for the Jupiter Orbit Insertion, 1995 187 kg for the Perijove Raise Maneuver, 1996

Major Computers:

CDS = Command and Data Subsystem
AACS = Attitude and Articulation Control Subsystem
The processors in the 8-bit computer ran at approximately 1.6 MHz, corresponding roughly to the microprocessors in the Apple II computers of the 1970s. Even a 486 would be about 200 times more powerful. But the Galileo computers were made for conditions in space—they were *much* less prone to crash than the typical PC.

Tape Recorder:

Capacity:	914,489,344 bits = 109 megabytes

Communication:

Radio frequency:	S-band, 2295.0 or 2296.5 Ghz
Broadcast power:	15 watts

Gain of the low-gain antenna (which had to be used because of the failure of the primary, high-gain antenna): +7 dBi (decibel over isotrope) when oriented ±10° toward Earth

Power reaching Earth from Jupiter: −197.5 dbW

Gain that the high-gain antenna would have had: +50 dBi in the X-band, +38 dBi in the S-band

Loss of transmission capacity because of the failure of the high-gain antenna: a factor of 20,000 (8 as opposed to 134,400 bits per second on arrival day)

Instruments: 11

SSI = Solid State Imaging Camera

Task:	mapping the Galilean moons with a resolution down to at least 1 kilometer and 20 months' monitoring of Jupiter's atmosphere (latter canceled due to failure of the high-gain antenna)

Details:

Median power needs:	17 watts	
CCD chip:	pixels:	800 × 800
	chip cooling:	to 163 kelvin (−110° Celsius)
Optics:	type:	Cassegrain
	focal length:	1,500 meters
	aperture:	176.5 mm
	focal length:	f/8.5
	field of view:	0.47° ½ full-Moon diameter)
Filters:	number:	8
	wavelengths:	380 mm to 1.1 μm
Exposure times:	2⅓, 8⅔, 30⅓ and 60⅔ seconds	
Dynamic:	8 bit	
Transmission time		
for one image:	with the high-gain antenna: one minute without the main antenna: several hours (with no data compression it would have been nine hours; with data compression typically two to three images a day were received)	
Mass:	18 kg	

NIMS = Near-Infrared Mapping Spectrometer

Task:	observing Jupiter and its moons in the infrared range in order to explore their atmospheres, and surfaces
Details:	spectral range: 700nm to 5.2 μm
Mass:	18 kg

UVS = Ultraviolet Spectrometer
Task: measuring the gases and aerosols in Jupiter's
                            atmosphere
Details: spectral range: 115 to 430 mm
Mass: 4 kg

EUV = Extreme Ultraviolet Spectrometer
Task: investigating the emission of ions in the Io torus
                            and the light phenomena in Jupiter's atmosphere
Details: wavelengths from 54 to 128 nm
Mass: 13 kg

PPR = Photopolarimeter-Radiometer
Task: investigating the distribution and types of atmos-
                            pheric particles and the surface reflectivity
Details: discrete bands in the visual and infrared to greater
                            than 42 μm
Mass: 5 kg

MAG = Magnetometer
Task: monitoring magnetic fields
Details: detection range from 32 to 16,384 gamma
Mass: 7 kg

EPD = Energetic Particle Detector
Task: detecting energetic electrons, protons, and heavy
                            ions in Jupiter's magnetosphere
Details: detection range for ions: 20 keV to 55 MeV; for
                            electrons: 15 keV to 11 MeV
Mass: 10 kg

PLS = Plasma Detector
Task: determining the composition, energy, and three-
                            dimensional distribution of electrons and low-
                            energy ions
Details: 64 bands from 1 eV to 50 keV
Mass: 13 kg

PWS = Plasma Wave System
Task: confirmation of electromagnetic waves and the in-
                            teraction between waves and particles
Details: detection range for E-waves: 5 Hz to 5.6 MHz; for
                            B-waves: 5 Hz to 160 kHz
Mass: 7 kg

DDS = Dust Detector System

Task:	measuring the mass, speed, and charge of dust particles
Details:	sensitive to 10^{-16} to 10^{-6} grams
Mass:	4 kg

RS = Radio Science

Task:	determining masses through the Doppler effect, and transmitting radio waves through atmospheres
Details:	uses the orbiter's normal S-band radio
Mass:	0 kg, since no extra hardware was needed

HIC = Heavy Ion Counter

Task:	detecting heavy ions
Details:	sensitive to elements from carbon to nickel
Mass:	8 kg

A motor held the lower part of the 4.5-meter-tall orbiter—the "despun section"—completely still. The solid-state imaging camera, with the same optical system as the Voyager camera plus a CCD (charge-coupled device) chip, the near-infrared mapping spectrometer (NIMS), the ultraviolet spectrometer (UVS), and the photopolarimeter/radiometer (PPR) were fixed to a moveable "scanning platform," which also carried a relay antenna for the probe's weak radio signals. One additional "instrument"—requiring no hardware, but used in practically all space probes—is called "radio science." The normal broadcast signal from the spacecraft's onboard transmitter is used as a tool. The tiniest changes in the speed of the spacecraft relative to Earth become evident in the form of frequency variations. This is the Doppler effect, according to which, in its well-known acoustic version, the horn of an approaching car sounds much higher than the same horn when the car is moving away. For the listener, the sounds waves are either compressed or stretched out, and the same thing happens to light and radio waves. Radio science uses these facts to determine the masses of planets and moons by measuring how much they deflect the probe from its course. It is also possible to detect concentrations of mass beneath the surface of celestial bodies, and to learn about their atmospheres and ionospheres by broadcasting radio waves through them.

Just how many onboard instruments there were on Galileo is ultimately a matter of definition. The most important ones are the nine mentioned above, plus radio science. Beyond these, however, the payload included two more items that are sometimes counted and sometimes not. One was a detector for extremely short wavelength ultraviolet radiation, which had proved its value as a supplement to the Ultraviolet Spectrometer (UVS) on the Voyager mission. The second was a device called the Heavy Ion Counter (HIC), designed to count the number of times particularly heavy ions collide with the spacecraft. This aspect of the mission was more an experiment thought up by the engineers than an actual part of the scientific payload. In the 1980s, it was popular to call Galileo the "Rolls Royce of space probes." Such a complicated instrument had never existed before. It was packed full of electronics, including 22 microprocessors and some 85,000 individual parts, corresponding to 46 million transistor functions. In technological complexity, Galileo stood far above its Voyager predecessors, even though the electronics they contained were already the equivalent of about 2,000 color television sets.

Primarily responsible for two of the Galileo instruments, one each on the orbiter and the atmospheric probe, were two "principal investigators" from Germany. Eberhard Grün was in charge of the dust detector, and Ulf von Zahn was responsible for the helium detector in the atmospheric probe. Other German scientists and institutions were involved in the development and operation of other instruments as well, including in the evaluation of camera data. Certain tasks while the probe was under way to Jupiter were also undertaken from the *Deutsches Zentrum für Luft- und Raumfahrt* (Center of Aerospace Technology) ground station in Weilheim, Germany. Scientists there were especially interested in radio science, as well as possibly getting a scan of the Sun's atmosphere, the corona, if Galileo's radio signal proved able to penetrate it. Galileo's engines also came from German workshops, although the complications with the engines mentioned above were still causing concern at the end of 1989, because uncertainty as to their effectiveness hampered the planning of the complex mission. It remained unclear whether it would be possible to approach both, or only one, of the asteroids. Without question, the priority remained on completing the mission in Jupiter orbit, with two years of maneuvers from moon to moon.

In preparation for the part of the mission that would start 50 days prior to Galileo's arrival at Jupiter in 1995, the rotation of the probe was to be accelerated to ten revolutions per minute—this time including the otherwise stable despun section, because the atmospheric probe was mounted underneath it. Spin-stabilized in this way, the probe would require no propulsion of its own to fly precisely toward Jupiter over the ensuing five months. The orbiter would continue on its course toward the moon Io, using its main engine to steer into Jupiter orbit in December 1995, while 214,000 kilometers below, the probe would just be descending into Jupiter's atmosphere. After undergoing a hair-raising braking maneuver, the instruments would descend through ever deeper atmospheric layers toward Jupiter, all the while transmitting data to the orbiter. Critical to this process was the Neutral Mass Spectrometer (NMS), which had the job of determining the composition of the atmosphere, and a special instrument designed to measure the relative abundance of helium and oxygen, the two most primitive elements. Plans were to take readings of temperature and atmospheric pressure and density (Atmospheric Structure Instrument, ASI), in addition to monitoring atmospheric particles (nephelometer, NEP), lightning (Lightning and radio Emissions Detector, LRD), and radiation flux at a variety of altitudes (Net Flux Radiometer, NFR). There was even a special instrument along, the Energetic Particles Instrument (EPI), to investigate the magnetosphere just over the top of Jupiter's clouds.

Galileo's Atmospheric Probe

Mass:	339 kg, of which 121 kg = descent module, of which 30 kg = scientific payload
Entry speed into Jupiter's atmosphere:	47 km per second = 170,700 km per hour
Instruments:	6

ASI = Atmospheric Structure Instrument

Purpose:	measuring pressure, temperature, and density as a function of altitude
Details:	range: 0 to 540 kelvin and 0 to 28 bars
Mass:	4 kg

NMS = Neutral Mass Spectrometer

Purpose:	determining the chemical composition of Jupiter's atmosphere
Details:	1 to 50 AMU (atomic mass units)
Mass:	11 kg

HAD = Helium Abundance Detector

Purpose:	measuring the abundance of helium in the atmosphere
Details:	precise to 0.1 percent
Mass:	1 kg

NEP = Nephelometer

Purpose:	confirming clouds and particle analysis
Details:	particle size: 0.2 to 20 μm; three per cm^3
Mass:	5 kg

NFR = Net Flux Radiometer

Purpose:	determining the radiation flux at different altitudes
Details:	six measurement ranges, from 300 nm to 100 μm
Mass:	3 kg

LRD/EPI = Lightning and Energetic Particles Investigation

Purpose:	observing storms and detecting energetic particles in the inner magnetosphere
Details:	sensor with fish-eye optics and radio sensor for 1 Hz to 100 kHz (LRD); sensors for electrons, protons, alpha particles, and heavy ions (EPI)
Mass:	2 kg

"Cruising" to Venus and Back

Meanwhile, it would be a good six years before data from these experiments would begin to arrive—if all went according to plan. For some of the instruments on board the orbiter, however, the acid test began early in Galileo's "cruise" to Venus and back, just a few days into the launch. EPD and HIC had already been turned on, and HIC had quickly detected a pair of ions that had just been released by a massive explosion on the Sun. Even the CCD chip in the camera had registered the impacts of a few isolated particles. Scientists anxiously awaited the first of what would amount to 30 major and minor course corrections in the period from November 9 to 11—and were relieved when it was made much more precisely than they had feared might be the case because of the thruster problem. Still, estimates regarding the fuel supply remained unchanged. There would be just enough for an encounter with Gaspra, but not for Gaspra and Ida both. For that, Galileo seemed to be a pathetic 30 kilograms short.

It was not all bad news. The Sun shield that had been installed on Galileo for its detour to Venus was functioning superbly, keeping the probe cool. The tiny auxiliary antenna was also working well. After the second course correction on December 22, 1989, preparations got under way for the Venus flyby, scheduled for February 9, 1990. A number of the instruments were now turned on permanently, but only in early January did Galileo's manager give the okay for experiments, announcing that observations of Venus posed no undue threat to the spacecraft, which would be enduring the greatest heat at precisely that time.

When Galileo made its closest approach to Venus on February 9 at 9:58 P.M. California time, it came just 18 seconds earlier than expected—that was the extent of the deviation. In spatial terms, the spacecraft had missed its target by only 5 kilometers, flying 16,106 kilometers over the Venusian cloud cover. The main objective of the

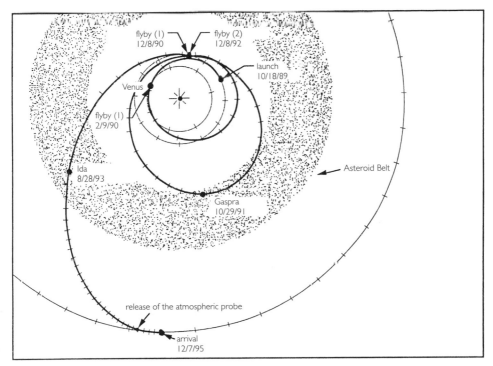

Planet-surfing to Jupiter: Time ticks on Galileo's trajectory and on the orbits of Venus and Earth are every 30 days, and on Jupiter's orbit every 100 days.

Venus swingby, of course, had been to gain kinetic energy, and the effort had paid off. The momentum Galileo picked up from Venus for nothing would have cost one and a half times its total supply of fuel.

The scientific program was also in full swing. The first 81 photographs that had been taken by the camera were now stored on tape. Flight controllers got to work immediately transmitting three selected images to Earth. Most would be sent in October over the low-gain antenna when Galileo came back into Earth's neighborhood. But getting the three images, plus a smattering of infrared data, required scientists to outsmart their own instruments. The software had been designed to transmit scientific data over the high-gain antenna on Galileo's later approach to Earth, but now there was only the small antenna installed specifically for Venus. Sending a single image took over four hours. But the quality was good, and the results from a spot check encouraging.

The big news, however, concerned the thrusters. Performance figures came in two percent better than the pessimistic estimates, drastically altering initial expectations of fuel supply. Instead of having 30 kilograms too little fuel, now there was a 20-kilogram surplus. Suddenly it seemed that the mission in its entirety might be possible. Galileo would alter its course for an encounter with both asteroids, and, once in orbit around Jupiter, it would pay a total of ten visits to the planet's moons. The targeted Gaspra flyby was put definitely on the flight plan, at least for the moment.

First, Galileo had to be steered back toward Earth for the initial rendezvous on December 8, 1990. The course correction on April 9 had already reduced the closest projected approach for Galileo from 2.4 million kilometers to 500,000, and later maneuvers would cut that distance to only 950 kilometers. The control jets would be fired 6,372 times to achieve the overall correction, burning 20 kilograms of fuel. But the hottest phase of the mission had been accomplished. In November 1990, while a series of small adjustments made the Earth flyby even more precise, all the stored data were transmitted. Even a quick preliminary look revealed surprises, although a full evaluation of the Venus observations, of course, would take time.

Learning About Earth's Sister Planet

One thing Galileo had not been able to do: it could not get a look at Venus itself through the thick clouds that permanently cover the surface of the approximately Earth-sized planet. That task was left to the Magellan orbiter, which arrived in August 1990 and since then had been transmitting a flood of images of the complicated volcanic structures that dominate the Venusian landscape. Nevertheless, the sole instruments on board the economical Magellan probe were radar and a radar-intensity meter, so that it was up to Galileo's battery of cameras, antennas, and detectors to clarify several important and previously unknown facets of the Venusian world. One of the biggest open questions was whether the planet's volcanic activity was ongoing or had reached its end. A vigorous dispute had been raging in the technical literature, centered on radio signals that the Pioneer Venus orbiter and a Soviet lander craft had supposedly "heard" repeatedly on the surface. At issue was not the radiation itself, but how to understand it. Did it come from electrical phenomena far above the Venusian surface, or was it lightning from storms?

Proof of the existence of storms on Venus would have increased the probability of ongoing volcanism. The Venusian atmosphere has none of the updrafts and downdrafts that are largely responsible on Earth for keeping positive and negative charges separate, without which lightning cannot occur. Volcanic activity would point to another mechanism for separating the charges. Lightning bolts have been observed in the clouds of ash ejected from volcanos on Earth, caused by the friction of the ash particles rubbing against each other, which produces an electrostatic charge. Variation over time in the amount of sulfuric acid droplets in the atmosphere would have been another indication of at least intermittent volcanic activity on Venus, but heated discussion was provoked nonetheless by the radio signature Magellan had picked up from the planet. The Plasma Wave System (PWS) on board Galileo was as sensitive to higher radio frequencies as any other detector that had ever visited the planet—and it had made a discovery. No better than any other instrument at determining whether the radio signals were produced by storms, PWS, during 53 minutes of continuous monitoring of the planet, had registered a total of nine unusual "impulse events," characterized by brief intensity spikes. Six of these, in the 1-megahertz region, could be interpreted only as the kind of lightning produced by stormy weather. Storms in the atmospheres of Saturn and Uranus had already been revealed in much the same way by instruments very similar to those on board Galileo. It is true the intensity of these six "events" had not been especially high, but Galileo had not come particularly close to Venus, either. Lightning of a comparable intensity was known to have been produced by storms on Earth, with lightning discharges on Venus perhaps even two to four times stronger. This time, even skeptics acknowledged the reality of the lightning—only to be proven wrong a decade later by the Cassini spacecraft on its way to Saturn. When it flew by Venus in 1998 and 1999, its much more sensitive radio instrument did not register a single lightning strike, while the same instrument had no trouble a few months later detecting lightning on Earth at the correct rate and intensity, all but throwing out the case for lightning on Venus.

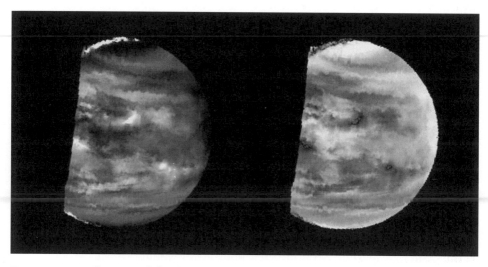

Two versions of a map of the deep clouds on the nighttime side of Venus, from readings taken by the Near-Infrared Mapping Spectrometer (NIMS) on February 10, 1990. At wavelengths of 2.3 μm the instrument could see to a depth of about 15 kilometers into the clouds covering Venus in the visible spectrum, where it identified another layer of clouds about 50 kilometers above the surface. On the left are the actual NIMS readings, with light areas indicating thermal currents in the deeper atmosphere and the surface of the planet appearing through holes in the cloud cover. On the right is a simulation based on the NIMS data, showing how Venus would look if the outer layer of clouds could be removed, allowing the layer at 50 kilometers altitude to be seen in reflected sunlight. Individual cloud structures are visible near the equator, while the clouds farther toward the poles are stretched into long thin shapes.

There was still more to be learned from Galileo's Venus data. In less than a week, 77 photographs had been taken, both in the violet (418 nanometers) and the near-infrared (986 nanometers) spectra, in which Venus had never been observed. The different colors make it possible to peer into the thick layers of clouds to different depths. It was very much what atmospheric scientists had in mind for Jupiter, with the difference that on the gas giant, unlike Venus, there was no solid surface underneath the clouds. Extensive research over 30 years had already been devoted to the structure of Venus's clouds, which had been photographed from a distance or even flown through by U.S. and Soviet planetary probes, and developments in ground-based infrared telescopes had also made new contributions possible.

To the unaided eye, however, the Venusian clouds present an impenetrable, radiant white ball, with virtually no clear structure. Only in the violet and ultraviolet does a distinct pattern of stripes become visible, with the contrast appearing at an altitude of 65 to 70 kilometers above the surface. Atmospheric pressure measures 50 millibars at that level, and the cloud temperature is about 40° Celsius. The clouds race around the planet, essentially parallel to the equator, at a speed of about 100 meters per second. The planetary surface underneath moves about 50 times more slowly, with this extreme degree of "superrotation" representing one of the big unanswered questions about Venus. Galileo's photography in the violet once again showed the familiar pattern of light and dark bands, forming a global spiral in the direction of the poles. Both the specialized Mariner 10 probe and the Pioneer orbiter in the 1970s had already sent back such images. The clouds obviously form in clumps at low latitudes and are then stretched out lengthwise by shear forces at the same time that they drift toward the poles. In the near-infrared, however, Galileo peered more deeply into the clouds, discovering a pattern—if of relatively low contrast compared with the images in the violet—that no one had ever seen before.

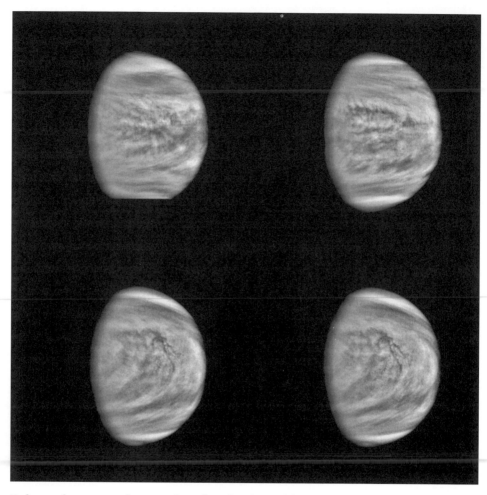

Enhanced-contrast photographs taken by the Galileo camera, offering an especially clear view of the cloud structure just one week after closest approach. Of particular interest are the complex patterns near the subsolar point, where heat from the rising Sun drives powerful convection currents.

In this perspective, the polar region appeared dark. Cloud density obviously increased rapidly at an altitude of about 60 to 65 kilometers. Also obvious at lower altitudes was the general spiral shape of the clouds, but there they were less sharply inclined toward the equator. And, the lower the altitude, as expected, the slower the clouds were moving, at 78 meters per second in the near-infrared, as opposed to 101 in the violet. Yet, the "windows" in the infrared wavelengths had still another advantage that could be exploited in ideal fashion by NIMS, Galileo's Near-Infrared Mapping Spectrometer, analyzing the radiation spectra deep in the atmosphere. NIMS confirmed what ground-based measurements had already suggested: the Venusian atmosphere is bone dry, containing even less water vapor than had been assumed. If that is the case, however, the greenhouse effect would no longer be adequate on its own to explain the high surface temperatures. The heat comes primarily from carbon dioxide, the main ingredient in the Venusian atmosphere. It allows radiation with short wavelengths in, while trapping longer wavelengths inside. Still, the greenhouse model called for a certain amount of moisture to reinforce the effect, and it simply was not there. Galileo had solved one of Venus's mysteries—leaving an even more fundamental puzzle behind. And the tour had just gotten under way . . . .

Return to Earth

On December 8, 1990, Galileo was almost home again, racing past Earth at only 961 kilometers altitude. The ground track, or the spacecraft's trajectory projected onto the Earth's surface, cut right across Asia, Africa, and the Atlantic Ocean, with the point of closest approach occurring over the Carribean Sea. There was reason to celebrate the very first evening. Galileo had missed its imaginary target point over the Earth by only 8 kilometers in space and a half-second in time, a major accomplishment for the navigation team. The exactness of NASA's predictions had even allowed an amateur astronomer in Texas to shoot a couple of snapshots of Galileo through his telescope that morning, when the probe was still 600,000 kilometers away. Again, it had picked up momentum—energy, strictly speaking, that our planet no longer has. Not that there is any great loss involved: after a billion years, Earth will be just 7 centimeters behind where it otherwise would be in its orbit around the Sun. There was more at stake, however, in Galileo's first Earth flyby than working out the mechanics of its flight course. A lot of science would also be done, in regard to both the Moon and the Earth itself. Project managers had already decided that a year earlier, as soon as it became clear that using the instruments now would pose no risk to their primary mission.

All Galileo could see of the Moon at first had already been explored by Apollo orbiters and astronauts. That gave technicians a good opportunity to run tests on Galileo's various cameras and spectrome-

Galileo captured unusual views of our own Moon during its 1990 flyby of the Earth–Moon system. The spacecraft's camera was able to see lunar regions that are normally turned away from Earth. The picture on the left shows the lava-filled "seas" known to Moon watchers, the Oceanus Procellarum and the Mare Imbrium, whereas the picture on the right is dominated by the great Orientale Basin on the Moon's nighttime side. Toward the bottom on the left, the very old Aitken Basin at the south pole can be made out. The nighttime side is completely dominated by cratered highlands, and the daytime side by the seas.

ters after a year in space, checking for any drift that might have occurred in their calibrations. Later Galileo also got a look at the Mare Orientale and the heavily cratered dark side of the Moon, which remained, after 30 years of lunar research, less well known than the side we always see. Galileo's camera, NIMS, UVS, and PPR would study the composition of the Moon's surface on the nighttime side, which scientists expected would differ from the side we normally see.

The distribution of water vapor in the Earth's mesosphere, one of the outer atmospheric layers, was among the main objectives of Galileo's research on our own blue planet. An attempt would be made with the EUV (Extreme Ultraviolet Spectrometer) to model the ionosphere in the helium and positively charged oxygen spectra, while the UVS would look for the geotail (the lengthening of the Earth's magnetosphere in the direction away from the Sun) in the light of hydrogen emissions and also determine the amount of ozone over Antarctica. The PPR, for the first time, would get a reading from space on how sunlight reflecting off Earth is polarized, and technicians also activated the particle detectors. The Earth's mass would be measured for the first time using radio science from a passing probe—something astronomers had already done with all the other planets from Mercury to Neptune.

The optical high point was yet to come. For years, NASA had been anticipating a series of color photographs Galileo would take of Earth—universally called "the Movie" by those involved in the mission—, promising views of the blue planet the likes of which no space mission had yet delivered. Data kept pouring in throughout December, not only regarding Earth and the Moon, but including all the information that had already been gathered about Venus. Galileo's perspective on Earth was indeed unusual, cutting diagonally up from the south with Australia, and especially Antarctica, as the most conspicuous attractions. The film version of the images was just as fascinating, with the reflection of the Sun moving across the ocean, spirals of

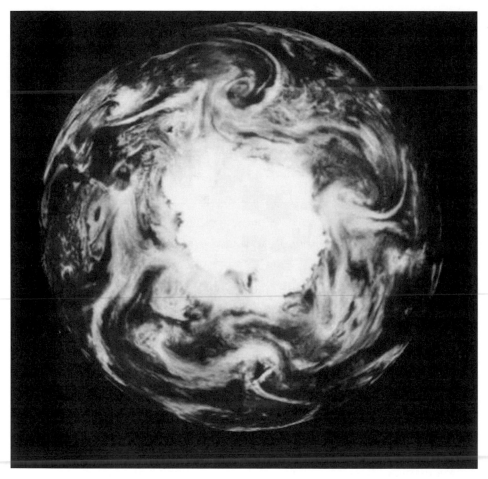

A picture that is actually impossible: the whole of Antarctica lit by the Sun. Since the Sun never shines on all of the gigantic continent at one time, several Galileo photographs taken over the course of 1 day had to be combined to produce this view, published for the first time in 1997. Taking such a picture had always been impossible before for another reason: nearly all satellites on polar orbits fly much too low to capture such a panoramic view.

clouds rotating, and the Earth slowly revolving. Galileo's movie of Earth quickly became part of the standard television repertoire. A private German broadcaster played a few seconds of the sequence for years afterward, as a lead-in to the evening news. The movie is also still used by the BBC as the background for some of its program notes.

Who is ever likely to have known they were seeing snapshots taken by a spacecraft headed for Jupiter? In addition to these images of Earth, Galileo also narrowed the sights of its camera and NIMS on the continent of Australia.

Galileo's lunar photography too is distinguished by its unusual perspective. The side it normally shows to us is dominated by extensive dark "seas," which in reality are lava flows. There are no such basins on the nighttime side, with the exception of the magnificent Mare Orientale, which was practically centered in Galileo's field of view. Even more important was the first good picture of the Aitken Basin, an impact crater 1,900 kilometers in diameter near the Moon's south pole. At some time an asteroid 150 kilometers across must have hit the surface here, possibly plunging deep into the iron-rich mantle of the Moon. Galileo photographs taken in various colors show clearly that iron is abundant in the Aitken Basin. The Orientale Basin, on the other hand, just over half as big, does not show any ferric accumulation—because there only the Moon's crust was disturbed by the impact of an asteroid. Galileo's photographs could be evaluated as classic examples of remote planetary research, giving scientists the chance for a test run before turning their attention to Jupiter. But Galileo also conducted one very novel experiment, the peculiar results of which were published three years later, answering the question of whether life exists on Earth. Probably it does, possibly even intelligent life, but the positive evidence Galileo gathered has nothing to do with the large-scale human constructions that are visible even from space.

Is There Life on Earth?

The real meaning of this curious "control experiment," inspired by who else but the astronomer Carl Sagan, was clear. How reliable would one of our space probes be in investigating the existence of life on

some *other* planet? Projected further into the future, the question concerned how easy, or difficult, it would be to use supertelescopes in
Earth orbit to search for life on planets revolving around other stars. In
the early 1990s, a research program along those lines actually appeared
on a list of NASA's long-term goals for the U.S. space program, to become a reality perhaps in the second decade of the twenty-first century.
Exobiologists of coming generations will certainly make use of the report "A Search for Life on Earth from the Galileo Spacecraft," published by Carl Sagan and his team in the October 21, 1993, issue of *Nature*. "At ranges varying from ~100 km to ~100,000 km, spacecraft
have now flown by more than 60 planets, satellites, comets and asteroids," the study says. "In none of these encounters has compelling,
even strongly suggestive, evidence for extraterrestrial life been found.
For the Moon, Venus and Mars, orbiter and lander observations confirm the conclusion from fly-by spacecraft. Still, extraterrestrial life, if
it exists, might be quite unlike the forms of life with which we are familiar, or present only marginally. The most elementary test of these
techniques—the detection of life on Earth by such an instrumented
fly-by spacecraft—had until recently, never been attempted."

Now, using NIMS, UVS, the camera, and the plasma wave spectrometer, Galileo had conducted just that experiment, requiring those
who analyzed the data to ignore all of their prior knowledge of the Earth
and its biosphere. "In what follows, we do not assume properties of life
otherwise known on Earth, but instead attempt to derive our conclusions from Galileo data and first principles alone." According to the fundamental principle of the search for life on other planets, any significant
deviation from thermodynamic equilibrium is a necessary, but not sufficient, condition for the existence of life. Had any such "disequilibrium"
been discovered—say, the presence of a specific chemical compound in
excess of expected values for chemical equilibrium by a factor of some
tens—the next step would be the systematic elimination of all nonbio-

logical explanations for it. "Life is the hypothesis of last resort," insist Sagan and his colleagues. The abundance of ozone in Earth's atmosphere, for example, exceeds the equilibrium by a factor of 10^{20}—the result of photochemical reactions driven exclusively by solar radiation.

One of Galileo's important "discoveries" about Earth was the presence of water in a variety of forms. NIMS spectra clearly identified condensed water in the Antarctic. Given that the temperature was pegged at $-30°$ Celsius (also correctly), it could only be ice. It was even possible to determine the existence of snowflakes from 0.05 to 0.2 millimeters in size. NIMS also concluded that large parts of the planet were covered by liquid water—what we call "oceans"—, that water vapor was present throughout the atmosphere, and that the atmosphere itself was full of oxygen. No world so rich in oxygen had ever been discovered anywhere in the Solar System. But where does the oxygen come from? And why, in contrast to Venus and Mars, is it still there? "In theory," there should be less oxygen remaining on Earth than on these other planets, because they have histories of new, oxidizable surfaces constantly being exposed to the atmosphere. An oxygen-rich atmosphere thus hints at the existence of life on Earth, but is very far from definitive proof.

Methane gas is a different matter altogether. It oxidizes so quickly into carbon dioxide and water that in equilibrium conditions there would not be a *single* methane molecule in the atmosphere. But methane is obviously present—in abundance, by a factor of an incredible 10^{140} relative to the thermodynamic equilibrium. Something must be constantly pumping methane into the atmosphere, at the rate of about 500 megatons a year. Now *that* is a rather strong indication of the presence of life, because no geological source could produce such quantities. This reading of the Galileo data by a "naive" technician would be correct, as we know, because roughly half of the methane in the atmosphere comes from methane-producing bacteria, while the other half is contributed by human sources, including agriculture and the burning of fossil fuels.

Interpretations of the camera images, despite being taken through a variety of filters, are once again ambiguous. Certainly, the large oceans are obvious from the way they reflect sunlight, as the NIMS data had already suggested. On dry land there are bright areas, looking like mineral basins, but also dark regions curiously green in color. The optical spectrum reflected by this "material" has unusual characteristics, which "cannot be uniquely interpreted" but which are in any case "inconsistent with any known rock or soil types on terrestrial planets" (Sagan et al.). Could the high absorption levels in the red part of the spectrum mean that a pigment of some sort exploits sunlight through photosynthesis? In that case, the oxygen-rich atmosphere would suddenly make sense! Our exobiologists have been growing more and more excited, and what we could tell them, of course, is that the mysterious green they have discovered is nothing other than chlorophyll.

Taking the exploration of suspected life on Earth a step further, our scientists might ask whether it had perhaps developed a civilization, leaving behind more clues that they could gather from afar. And this, they would simply deny—it was their bad luck that the best photographs, with a spatial resolution down to 1 or 2 kilometers, showed only the deserts of Australia, the least populated regions on the planet. Early rumors that a research station on Antarctica had been photographed successfully also failed to stand up. "No reworking of the surface into geometric patterns, or other compelling indications of artifacts of a technological civilization could be discerned," the analysts conclude drily. But they were hardly breaking new ground. Thirty years earlier, evaluations of countless images collected from orbit with a resolution of 1 kilometer turned up only a very few signs of life. Perhaps Galileo, flying past the dark side of Earth, would have seen the lights of cities, but no such photographs were possible, and, given the nature of the experiment, there could be no chance to conduct it a second time. The point, after all, was to simulate the results of a single flyby.

Still, hope remained for humanity. Galileo's plasma wave instrument picked up a strange emission in the narrow radio band of 4 to 5 megahertz for which there was no ready natural explanation. These isolated radio signals constituted "the only indication of intelligent, technological life on Earth" (Sagan et al.). They appeared only on the nighttime side of the planet, where the ionosphere is less disruptive to these wavelengths than it is by day, suggesting that they originated beneath the ionosphere. The broadcast source had to be located either in the atmosphere or on the planet's surface. The frequency of the signals remained constant for long periods of time, a strong indication that they came from an artificial source, because natural radio emissions are practically always characterized by a drift in frequency over time. Most impressive of all, however, was that the signals' amplitude was modulated in a way that the low temporal resolution of the instrument made it impossible to decipher. This, of course, strongly suggested an artificial source and, indeed, that information was being transmitted.

Great quantities of water and oxygen, the "mysterious" red-eating pigment, and the excess of methane were thus the only indications of life forms. Yet, Galileo could have taken the same readings 2 billion years ago, when only "primitive" life forms populated the Earth's lands and seas. All that could possibly be regarded as evidence of intelligent life was the radio signals. Now, "life as we know it" exists in a state of powerful reciprocal interaction with its environment, and so would betray itself to Galileo and similar visitors. Yet it is also possible to imagine plausible ecosystems—subterranean ones, for example—that would be much more difficult to detect. For Mars and, as we shall see later, for Jupiter's moon Europa, such scenarios of hidden life are even very plausible. In any case, proper scientific method is still being used to evaluate the hypothesis that there might be underground inhabitants on Mars or "Europeans" living beneath the ice on one of Jupiter's moons. Thus Carl Sagan's final conclusion on Galileo's unusual experiment is this: "Our results are consistent with the hypothesis that widespread biological activity now exists, of all the worlds in this Solar System, only on Earth." The existence of biological life forms elsewhere, even on our neighboring planets in the Solar System, in other words, must by no means be ruled out.

The Darkest Hour

After the successful rendezvous with Earth in December 1990, the next item on Galileo's itinerary was Gaspra. For the first time, a spacecraft was to pass by one of the thousands of asteroids populating the Solar System in between the orbits of Mars and Jupiter. A series of course corrections began on March 20, 1991, in preparation for the October 29 encounter. Then came a flurry of activity on March 26, when one of the most important onboard computers shut itself down because of a systems error. The incident precipitated a so-called safing event, when the scientific instruments were turned off and the bottom portion of the spacecraft, the despun section, was allowed to spin. No lasting damage was sustained, however, and soon Galileo was brought back under full control—until April, when a completely unrelated event would write the Galileo project permanently into the history books.

On April 11, 1991, according to the flight plan, Galileo's 4.8-meter high-gain antenna, which had been kept folded up under a heat shield to protect it from the Sun, was finally supposed to open. Flight controllers issued the command for the antenna to open—but it became stuck only halfway through what was supposed to be a routine ten-minute procedure. The mood at JPL was one of shock and confusion. Antennas of the same type had been used on a number of communications satellites and had always functioned flawlessly, so nothing of the sort had been anticipated. On board Galileo were hardly any systems capable even of analyzing the antenna's condition directly.

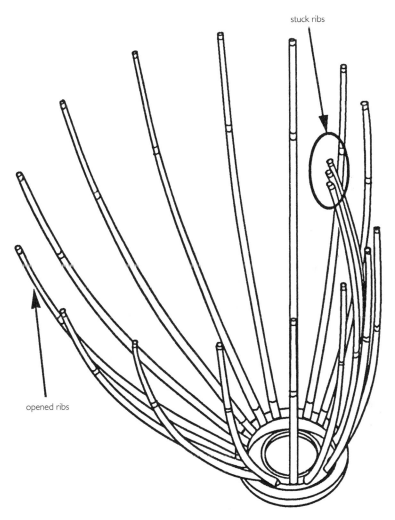

Condition of the Galileo antenna according to an analysis of the telemetric data: three of the ribs failed to disengage from the central mast (not shown here), making it impossible for the parabolic antenna to unfold and rendering it completely useless.

Only one thing was certain: it would not open. A microswitch that was supposed to notify controllers that the antenna was open had not done so, and the rate at which Galileo was rotating also had not slowed appropriately. Within a week it became clear that one side of the antenna had opened further than the other. The antenna itself, made out of gold-plated molybdenum wire, was stretched between 18 graphite-epoxite ribs. At launch time, it had looked like a folded-up umbrella, with closely spaced ribs fixed to a central mast. It was supposed to be opened by a motor (or its redundant backup) that activated a lever mechanism to extend the ribs. In principle, quite simple. But after just 17 seconds, the primary motor began laboring and finally stopped.

That could only mean that the ends of some of the ribs, which were supposed to be attached loosely to the central shaft, must have gotten stuck. Experiments with the reserve antenna, which had remained behind at JPL, revealed a similar problem. Now, the important thing was to find a solution. The first idea derived all but automatically from the jammed rigs. Galileo would have to be turned toward the Sun in such a way that the temperature change on the probe itself would deform the antenna. That should have the effect of allowing all the ribs to snap free—or at least that is how things went in the vague theory put forward in the name of managing a successful repair.

What would happen if the antenna could not be saved? The encounter with the asteroid Gaspra could proceed as planned, with the data transmitted over the tested low-gain antenna. For a meaningful mission to Jupiter, however—as believed universally at the end of April—the low-gain antenna would be totally inadequate. Some were already calling for the construction of a relay satellite that would be sent to Jupiter on a faster route, where it would orbit the planet with Galileo. Outfitted with two large antennas, the relay would be able to pick up Galileo's weak signal and transmit it on to Earth. Even more desperate was the suggestion made by others that everyone just forget the Jupiter mission and turn Galileo into a pure asteroid probe. It would maneuver over and over between these small celestial objects and then return to Earth to deliver the data.

Engineers at JPL, fortunately, did not give up so easily. This would not be the first time that a thorough analysis had allowed them to correct an apparently fatal problem on board a distant interplanetary probe. The jammed camera platform on Voyager 2 was one example. Had they not fixed that, much less data would have been gained from Uranus and Neptune. For Galileo, alongside the hopes invested in thermal deformation, there was also the possibility of starting both motors in little jolts and in that way slowly hammering the ribs loose. Yet another idea was to give the whole probe a gentle, but determined, shaking. It would undergo additional shocks in the time to come, when the atmospheric probe was released, for example, or when the powerful retropulsion module was fired to slow down the spacecraft for its approach to Jupiter. Perhaps those would help. A "tiger team," as working groups are called at JPL when they are assembled quickly to cope with an emergency, got to work. For the moment, the antenna could stay as it was, and the motors had no reverse gear in any case. Then on May 3, Galileo went into safing mode again, although this time the problem was quickly resolved. (It was not related to the problem with the antenna.) Meanwhile, the directive had come down that at least six months would be spent trying to get the antenna to open, before more radical measures were to be given any thought.

The best hope, as everyone knew, lay in the managed alterations of Galileo's temperature, although there was no agreement on which would help more, the heat from the Sun on the second Earth flyby or the cold of the Asteroid Belt beyond Mars. The answer emerged soon enough, when the first deliberate efforts to save the antenna, carried out from May 20 to 23, ended in utter failure. Galileo was rotated 45° around its axis, but since the Sun was so far away, what warming took place did absolutely nothing for the ribs.

A month passed since the problem had appeared before even a rough estimate of what had gone wrong became possible. Three, four, or five of the ribs had failed to detach themselves from the central mast, where they had been fixed at launch for reasons of safety. The cause of this seemed to be that an important lubricant had evaporated, either during the long period the probe was in storage or during one of the trips across the country between JPL and the Kennedy Space Center. On top of that a slight shrinkage of the central mast in the cold of outer space led to the ribs getting stuck. Because of the extreme fragility of the lightweight antenna, it had never undergone comprehensive thermal testing.

Another month later, the understanding at JPL of the antenna's unfortunate state was nearly complete. It turned out that only two of the 18 ribs were actually stuck, but in the near vacuum of space they were stuck all the more tightly. During all the cross-country traveling, the ribs had rubbed constantly against the mast, which is what caused the loss of the lubricant. Incidents of this sort were scarcely unusual, but no one had foreseen the danger.

Despite a public show of optimism, JPL was slowly running out of solutions. The first intentional cooling of the probe in July 1991 had also accomplished nothing, whereupon scientists planned an even more severe "cold soak" in August. They allowed Galileo to freeze for a full 50 hours, with a few of its components falling to −140° Celsius. The central shaft needed to shrink only 2 to 3 millimeters to release the two ribs, but again nothing happened. Now the last hopes lay with an even colder treatment. Galileo would be allowed to freeze to −170° Celsius.

Meanwhile, final preparations were being made for the encounter with Gaspra. Minor course corrections would steer Galileo to precisely the desired distance from the asteroid, 1,600 kilometers. However people may have been regarding the antenna crisis, everything possible was done to make this encounter a success.

Gaspra: The First Encounter with an Asteroid

Under way to its next destination, Galileo was practically mute. The onboard computer had already been given over completely to the task of intermediate data storage, so even if the antenna had suddenly opened, there would have been no time to reprogram it for "live broadcast." Moreover, the failure of the antenna, if that is what was happening, would make navigating Galileo much more difficult. The spacecraft would need photographs of Gaspra against the background stars to make its precisely targeted approach. These images could be transmitted only in fragments—four images at once, to be precise, rather than the planned 40. Navigation required flight controllers not only to locate Gaspra, but also to orient the camera platform extremely precisely—the camera, after all, had a field of vision about the size of the full Moon. So as not to miss the asteroid altogether, the camera was programmed to photograph a certain part of its "sky" systematically as the spacecraft sped by on October 29, 1991, at 8 kilometers per second, 1,600 kilometers from the primary target. Gaspra would be certain to appear in at least one of the frames. Initially there was thought of having a picture or two sent to Earth immediately after the flyby, but such plans were abandoned because of the extremely long transmission time involved. It appeared that the world would have to wait for a closeup of Gaspra until the second Earth encounter in December 1992.

All that was really known about Gaspra was that it was a class S asteroid, about $10 \times 11 \times 18$ kilometers in size, and that it made one rotation every seven hours. A midget compared with Amphitrite, a 200-kilometer boulder that a 1986 launch would have allowed Galileo to visit. One frustrated asteroid researcher even advanced the view that Gaspra had absolutely nothing of interest to offer. But most people saw it otherwise. "This mission is comparable to Columbus landing in the New World," rejoiced Clark Chapman, for example, anticipating "a remarkable first look at a completely new class of objects." On October 30, it became clear that everything had worked. The camera shutter had snapped 150 times and everything had been recorded on tape, not only images but also data from nearly all of the other instruments. Not only had Gaspra's surface been documented, but also its immediate interplanetary environment. Gaspra had never been suspected as a source of gas or dust, such as a comet gives off, but you could never be sure . . . .

Navigation, despite the difficulties, went even better than expected, as the final image fragments proved. There was a 99-percent probability that Gaspra would appear from the closest possible vantage point on at least one of the final images. For technical reasons, no pictures had been possible at the moment of the closest encounter. But photographs were taken of the slowly rotating asteroid during the hours, and especially the last half hour, of Galileo's approach. Only one shot was taken at a time at first, but then during the last phase a long succession of photographs was taken of the asteroid's most likely location. Now it seemed practically certain that Gaspra had been caught during this phase of the flyby, when the highest resolution was possible.

Naturally, the world wanted to see the picture right away—a desire taken by JPL as a challenge. Had flight controllers navigated precisely enough to make it possible to predict the individual frame in which Gaspra would appear? And could they dare monopolize the 70-meter antenna of the Deep Space Network for 80 hours transmitting at a mere 40 bits per second?

A week later the decision was made. They would go after the picture! Scientists decided not to transfer any of the sharpest images from about 5,300 kilometers away, because it remained impossible to predict which of these 51 shots would be the right one. But it did seem possible to locate Gaspra in the previous series of nine image fields. On November 14, 1991, it happened. Humanity got its first look at a world just as strange as it is tiny. Navigation turned out to be of unparalleled precision. The desired picture was taken between 35 and 29 minutes before the closest encounter. Getting the entire picture to Earth would have consumed an enormous amount of time, so at first only 12 lines were transmitted. And there it was—part of the asteroid could actually be seen! Now there was no stopping. Even so, it would be another four days before we would have our first look at Gaspra as a whole, a green-filter photograph taken from about 16,000 kilometers.

Encounter with an Asteroid

Gaspra was just slightly bigger than indirect astronomical readings had predicted, about 20 × 12 × 11 kilometers. And it was anything but a complete ellipsoid, lacking two big chunks—big dents were visible on the edges. Obviously, this celestial body had already been involved in a few cosmic collisions. Many impact craters were to be seen on the first image, but on the whole Gaspra's contours were soft, lacking in sharp edges and corners. What would that mean? In theory, all the debris from a collision should escape the asteroid's negligible gravitational field, but—and here the JPL image analysts were speculating freely—maybe always just a bit fell back to Gaspra's surface, leaving it with a consistency similar to that of our Moon. The finely ground soft stone mass on the lunar surface is called regolith. It is the product of billions of years of meteorite collisions, and (contrary to the fears of some), compacted enough that the Apollo capsule did not sink into it. Could this same process be at work on an object 200 times smaller?

Transmitting the first image in the green had worked so well that scientists immediately ordered the data from the other three color components. Haste was of the essence, because the data transmission rate was steadily slowing. In April 1992, Galileo already would have crossed beyond a critical distance. But by mid-November, one image in the violet and two in the infrared were actually there. Analysts put together the first color photograph, which might make it possible to draw conclusions about the mineralogy of Gaspra's surface by interpreting brightness variations in the different colors. The fundamental

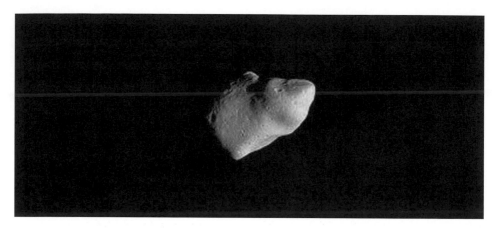

The first picture of Gaspra to be shown to the public. It was taken from a distance of 16,200 kilometers, with the illuminated part of the asteroid measuring 16 × 12 kilometers.

characteristics were already known from telescopic observation. While Gaspra was indeed a member of the S class of asteroids, it possessed higher than usual levels of olivine relative to pyroxene. And that said a lot about its past. It came from a much larger and more differentiated "progenitor body" that had been broken into pieces by a collision. Before the cosmic traffic accident, Gaspra had resided deep inside the larger body, close to the border between the core and the mantle.

However, the color photograph available in late 1991 was not all that exciting. Gaspra looked monotonously brown, and only with enhanced color contrasts did it become noticeable that the material thrown up from what were obviously fresh meteorite craters had a more bluish coloration. Finally, in June 1992, the public got a much sharper view of Gaspra. The transmission of two more images, taken from a distance of only 5,300 kilometers, went effortlessly in May and June. Together they showed the entire asteroid, in three times the detail of the first test sample from 1991. Instead of 165 meters, each

CCD pixel now corresponded to only 54 meters. And the number of recognizable impact craters rose to more than 600. Planetary geologists both in the U.S. and at the Berlin Center of Aerospace Technology got busy drawing curves, plotting the number of craters against their size. The plots could be used to test a variety of historical models of cratering in the Solar System—when did what bodies collide with which planets and where had they come from. Gaspra's age could also be determined, or more precisely, the time since it had been broken loose from inside its progenitor. The answer? At least 250 million years, but perhaps much longer, since counting craters is not an exact science.

In higher resolution images, Gaspra came into its own as a distinct world, with more than just the craters for a "landscape." Surface grooves became evident, for example—a phenomenon otherwise known only from Phobos, one of the Martian moons. The grooves on Phobos presumably come from stresses to which the small body is subjected. In Gaspra's case, they might be unhealed fractures left over from the breakup of the progenitor body. The result of this violent past is that Gaspra is one of the most irregularly formed bodies in the Solar System, deviating much more from a spherical shape than the Martian moons Phobos and Deimos, for example—and it may be that they in fact are asteroids caught in a Mars orbit. The "weathering" of Gaspra's craters was evident on the sharper images. Some were still quite sharp, with bright edges, while others (because of regolith?) had already taken on very soft contours.

Then again, maybe the truth is something else altogether, something much more complicated. By the time of the 1992 conference of planetary scientists and Gaspra specialists in Munich, a multitude of competing interpretations of the pictures had already been advanced. These meetings have been among the world's most fertile breeding grounds for new theories about the Solar System.

Big Puzzles on a Small World

Gaspra's craters and their size distribution require a bit more attention. No one disputed that it was a markedly "steep" distribution, with small craters greatly outnumbering big ones. The crater pattern on Gaspra bears rather strong similarities to the kind of cratering we see on our own Moon or on the planet Mercury, but it is very different from that which occurs on Phobos, where the number of small craters is much lower. Meanwhile, with experts still arguing over what Gaspra's craters say about the history of the Solar System, such comparisons were being called seriously into question. For the first time, scientists were dealing closely with an extremely small body left cratered by cosmic bombardment—and who was to say for sure that it would react in exactly the same way as the Moon or Mercury? According to new and finely detailed computer simulations of the asteroid (now established, after all the images had been analyzed, at 19 × 12 × 12 kilometers), the opposite is the case.

According to these new hydrodynamic calculations, any largish impact sends a shock wave running all the way through the asteroid, most of the energy of which goes to shattering the interior. The inside of Gaspra, on this model, would be a pile of rubble. On the other hand, relatively little energy remains to cast debris from the collision off into space, making it possible for larger craters to form than predicted by traditional analyses—craters as much as 8 kilometers across. Another important effect had also been overlooked. After every major impact, the entire surface of the asteroid "hops" several meters outwards, instantly erasing smaller craters. Whatever there is to be seen now on Gaspra that is less than a kilometer in diameter can only have happened since the last big collision, which left behind a crater several kilometers across. All this means, of course, that counting craters might not be a reliable or even possible way to calculate the age of such bodies. And a third important factor had to be considered in regard to crater distribution:

larger craters simply cover over their smaller predecessors, so that the evidence of a great many craters, especially the ones 1 to 4 kilometers in diameter, is lost.

The next task for image analysts was to find much larger craters, but their efforts were frustrated. Much of Gaspra's surface was dominated by extensive depressions, however, and they could in part be remnants of enormous collisions which, in theory, had erased the largish craters and set the "clock" otherwise provided by the smaller ones back to zero. Astronomers were stunned by this news—that the reliability of crater statistics on small bodies in the Solar System had been fundamentally cast into question. Disputes over what the surface depressions meant continued to appear years later in technical journals, sometimes directly advancing the heretical view of this tiny world and sometimes not. Whether these giant craters—at least 4 and possibly 8 kilometers in diameter—existed on Gaspra or not—making 3 kilometers the maximum crater diameter—remained open. Nevertheless, the theory that major collisions had distorted the crater evidence on this small asteroid, and that at least once the clock had been set all the way back to zero—that position ultimately found widespread acceptance. And it is generally acknowledged today that these cosmic midgets react differently to impact stresses than larger asteroids or full-grown planets. At this point, it remains unknown whether Gaspra came from inside a compact body or is simply a "rubble heap," a collection of individual rocks held loosely together by the force of their own gravity.

The last batch of Gaspra data arrived in December 1992, when Galileo reapproached Earth and relatively high rates of data transmission became possible once again. Now scientists could see the asteroid go through a complete rotation. Gaspra had appeared in a total of 57 of Galileo's 150 photographs, and roughly 80 percent of its surface was now known. A look at the other side of the asteroid yielded no immediate new insights. The distribution of craters was similar, and it remained impossi-

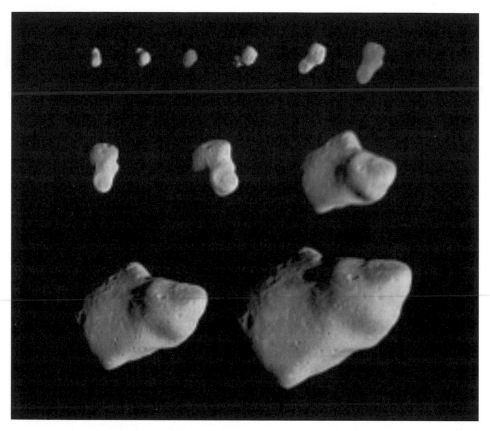

Gaspra times 11: Approach sequence shot beginning at 164,000 kilometers away and ending at only 16,000. The series lasted 5¾ hours, capturing almost one complete rotation, since it takes Gaspra about seven hours to turn once on its axis. The asteroid's "nose," pointing up in the initial image, first moves into the shadows before reappearing on the bottom left. Gaspra's "other side" is also free of large craters, at least of those on which all the experts can agree.

ble to discover any distinct craters more than 2 kilometers in diameter. Gaspra's size (18.2 × 10.5 × 8.9 kilometers) could now be determined exactly, giving it a mean diameter of 12.2 kilometers, plus or minus 0.8 kilometers. A view of the whole asteroid did reveal several astonishingly flat "plains," up to 7 kilometers in size but with a depth of less than 200 meters. Presumably, they were caused when Gaspra split off from its progenitor. After five years of crater counting, 200 million years remains the best estimate of the asteroid's life on its own in the universe.

Gaspra made headlines once again at the end of 1992. Among the last stragglers of data was the record of Galileo's magnetometer—and it was remarkable. At one minute before and two minutes after the closest approach to the asteroid, the direction of the interplanetary magnetic field reversed dramatically, without showing any change in intensity. At first, the magnetometer registered the asteroid as the focus of attraction, but then three minutes later the reading diverged sharply, finally to return to the original orientation. What had happened?

Galileo may have picked up a disturbance in the interplanetary magnetic field caused by Gaspra, exactly like what happens, though of course on a much smaller scale, near Venus, Earth, Jupiter, or near comets. To be able to cause the effect, Gaspra would have to possess a magnetic moment of its own, about 10 orders of magnitude smaller than Earth's. In planetary terms, that is almost nothing, but for a midget like Gaspra it was enormous. On its surface, the magnetic field would be as strong as the one on Earth, so a normal compass would work there! Gaspra is obviously much richer in metals than the regolith-covered surface had initially indicated. In July 1993, when this analysis of the magnetometer data was published, Galileo was already making its approach to Ida. A year earlier, given the mass of new knowledge gained from little Gaspra, NASA had decided to make the detour over to Ida, which was nearly twice as big. It would not be long before asteroid researchers had more to talk about than Gaspra.

Scenarios for Living Without the Main Antenna

Galileo had been so economical with fuel during the first two-and-a-half years of its journey that there were now plenty of reserves on hand, so dipping back into the Asteroid Belt in August 1993 posed no threat to the later tour of the Jupiter system. The scientists' experience with Gaspra had proved that it was worth taking a look at a second asteroid, a class of celestial objects that had barely been studied. The great success of the first encounter in the fall of 1991—when Galileo had opened up a truly new world for the first time—had also fortified the Galileo team in its belief that scientific discoveries about Jupiter were possible, if need be, without the high-gain antenna.

At that time a certain confidence remained that the antenna would eventually open. Additional deep-freeze treatments had been planned for December and the first half of 1992, in hopes of gradually freeing the ribs. Yet, it had been a year since the first attempt to open the antenna and there was still no movement. Switching the antenna motor quickly on and off had accomplished nothing, and now the Galileo team managers confided to the public that perhaps it would never work. Even in the worst-case scenario, "we can conduct a really exciting mission," promised project scientist Torrence Johnson, "much more spectacular than Voyager." Efforts to release the antenna would continue, he said, but eventually—if necessary—JPL would get to work developing a completely new Jupiter mission plan. About 2,000 pictures could still be transmitted with the low-gain antenna. That was 25 times fewer than planned, but with a careful selection of shots, it seemed possible that three-fourths of all mission objectives could be completed. When the new higher resolution Gaspra images became available in June 1992, NASA presented them to the public along with a definite plan. One last heroic effort would be made to open the antenna. Just after Galileo's second Earth flyby in December 1992, the antenna motor would be switched on and off a thousand times. Perhaps the antenna would be "hammered" loose.

Barring success, NASA said it would begin to develop a completely new plan for the Jupiter mission in March 1993, relying exclusively on the low-gain antenna. Over and over JPL team members made the point that scientific gains do not necessarily depend on the sheer number of images a spacecraft delivers. The Voyagers had transmitted some 100,000 pictures together, but all of the great discoveries were based on just a few thousand.

Off to Jupiter!

After the Gaspra visit, Galileo needed to pick up momentum one more time to make it all the way to Jupiter. That was to be taken care of with a particularly close Earth flyby on December 8, 1992. In early August, flight controllers made the first of several course corrections in anticipation of the flyby. The thrusters had been fired now a total of 5,400 times in the mandatory pulse mode—and fuel consumption, despite early fears, had risen only marginally. This is what had made it possible to include the second asteroid encounter on the itinerary. Like Galileo's first visit back to Earth, its final flyby exactly two years later would also be exploited for experiments. Galileo's instruments, aloft in space for three years now, would be calibrated a second time, enabling scientists to make very precise estimates of how much the sensors were drifting, which would make it possible to fine-tune final recalibrations at Jupiter. And there was more Earth research to come. Earth's own magnetosphere—in concrete terms, the "geotail" that trails Earth in the solar wind—would get some of the attention, as would the little-studied north pole region of the Moon. The target this time on Earth was the Andes Mountains, with hopes for some especially high-resolution pictures. The good results from Gaspra had proved that adjustments being made on the camera platform were working.

Since much is already known about the Andes from geological expeditions, scientists could put the camera and NIMS to a very precise test. Other observations of Earth were also planned, including an

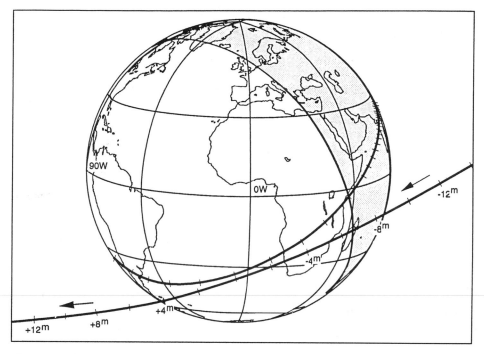

The second Earth flyby. Galileo's trajectory through space and its projection onto Earth's surface. Time ticks at two minutes.

ozone evaluation over Antarctica with NIMS, and once again, of course, a spectacular film would be created. Eight days after the flyby, the Moon, from Galileo's perspective, would be narrowly slipping by Earth. The camera alone would take a total of 6,813 pictures—more than could be expected now, with the defective antenna, from the entire Jupiter mission. An unusual experiment had also been planned. Ground-based telescopes would aim laser beams at Galileo in a first-ever test of "laser communications" with a free-flying spacecraft. Still, the most important aspect of the second rendezvous with Earth was clearly picking up the gravity assist for the trip to Jupiter. Galileo's heliocentric speed would be raised from 35 to the critical 39 kilometers per second. In purely mathematical terms, it would have been possible to attain the same velocity on a single flyby—but to do that Galileo would have to pass 1,600 kilometers closer to the Earth's center, which is far below the surface . . . .

The Second Return to Earth

Once again all went well. On December 8, 1992, Galileo darted by, this time 303 kilometers over the southern Atlantic. Flight controllers missed the space-time target by scarcely a kilometer and only a tenth of a second! Precision of that sort exceeded all expectations. It would make it possible to skip at least one major course correction, adding 5 kilograms to fuel reserves, which would bring the surplus up to 20—a good security cushion for the five years, minimum, that astronomers had planned for Galileo. From now on the spacecraft would be moving farther and farther away from the Sun, getting colder and colder. This meant that now, in the period between October 1992 and March 1993, would be the last chance for one final try to open the high-gain antenna. A total of seven thermic cycles had done nothing (aside from waste fuel by the kilo), and now all hopes rested on the "hammering" strategy. Ground tests with an identical reserve antenna had meanwhile demonstrated that the motor generated more force when it was switched on and off in pulses, rather than running continuously. For weeks the hammering went on, and at first the antenna indeed did move a bit. Then, very soon, there was no more progress, and by mid-January 1993 hopes were falling fast. After 13,320 pulses, on January 19 efforts were called off. Out of solutions, JPL turned to the task of planning a mission relying exclusively on the low-gain antenna.

Abandoning the primary antenna could not help but overshadow somewhat the satisfaction JPL was otherwise finding in the scientific gains of the second Earth flyby. The photographs of the Moon were so

Northeast Africa and the Arabian Peninsula as seen by Galileo's camera from about 500,000 kilometers on December 9, 1992. On the left are Egypt and the Nile Valley, in the middle Khartoum and the confluence of the Blue Nile and the White Nile, and at the bottom right Somalia.

spectacular that the camera team of the German Center of Aerospace Technology held its own press conference in Oberpfaffenhofen, and for what was probably the first time in the history of spaceflight, presented pictures from an American spacecraft to the German public prior to U.S. publication. The laser experiment was a great success as well. Regardless of whether the target was 600,000 or 6 million kilometers away, there was no difficulty hitting it with two laser beams from Earth. The feat was even performed on camera. Engineers celebrated "a milestone in space communications," having shown that it is possible to aim a laser at an interplanetary spacecraft solely on the basis of its predicted location. Future spacecraft would be able to communicate with Earth in this way—using 10- to 50-centimeter optical telescopes in place of the meter-size radio reflectors they now have on board.

Meanwhile, Galileo's investigations were by no means restricted to celestial bodies and actual light sources. To research scientists, the long journeys through interplanetary space present opportunities as well. This interest culminated in the spring of 1993 in a very unusual experiment that linked Galileo, the solar probe Ulysses, and the Mars Observer together to conduct a search for gravitational waves. Using the antennas of the Deep Space Network, physicists broadcast radio signals at precisely known frequencies to all three spacecraft, and each one transmitted the signals back. The predicted effect was a shift in frequency as a result of the Doppler effect, which would be caused by the motion of the spacecraft relative to the motion of Earth. However, if a very strong gravitational wave with an especially long wavelength were to pass through the Solar System during the time the measurements were being taken, scientists would be able to register a small but characteristic additional effect. A gravitational wave has never before been detected directly, but the phenomenon is integral to the general theory of relativity. Gravitational waves have eluded even the most advanced detectors, and this 1993 experiment failed as well. There had been no serious expectations that it would succeed, however. For a detectable wave to be produced, a cosmic catastrophe would have to happen somewhere in the vicinity of the Solar System—something on the order of a supernova exploding, for example, or two black holes colliding. Still, the experiment was worthwhile. Future research will probably rely on a system of space-based gravitational wave detectors. Widely separated spacecraft will remain in constant laser contact, maintaining an extremely precise record of their distance from each other.

No gravitational waves were to be had, but in June 1994, Galileo did run across an interplanetary phenomenon that had not been predicted. It is dusty in interplanetary space! This is not the interplanetary dust that has long been known, meaning the host of particles that revolve around the Sun with the major and minor planets, on the lower end of the planetary size scale, so to speak. Beyond 2 AU from the Sun things get more complicated. And they become extremely complicated near Jupiter. Ulysses had just sped past Jupiter in February 1992, outfitted with the same dust detector Galileo has on board. Normally, in interplanetary space, these detectors register only a single hit every one to ten days. Within a few months on either side of the Jupiter flyby, however, Ulysses had registered about ten bursts in the rate at which small dust particles collided with the probe. And these bursts appeared with a remarkable periodicity of 28 days, plus or minus three days. It was a completely new phenomenon!

The dust particles appeared in tightly bundled groups called dust streams. They were nearly all moving faster than 26 kilometers per second. And they must have come from the immediate vicinity of Jupiter. What was emitting the particles? What accelerated them? How was the periodicity to be explained? The second question was the easiest to answer. Electromagnetic forces were accelerating them. It had been expected that small dust particles would pick up charges from the plasma in Jupiter's magnetosphere and then be driven away from the planet by its magnetic field. No more could be known from Ulysses's quick look. But Galileo was going to spend years orbiting Jupiter and could examine the dust situation in detail.

Theorists were still puzzling over the electron mechanics of the Jovian dust streams when the Solar System prepared another surprise, in which Galileo would once again play a key role. On May 22, 1993, for the first time it became evident that in about a year, a comet might crash into Jupiter. P/Shoemaker-Levy 9, as the comet was designated technically, had been discovered only two months previously. It had caused a stir from the start, because it consisted of an entire chain of mini-comets. Soon mathematicians figured out that it had gotten too close to Jupiter in 1992 and had been torn apart by tidal forces. Enough data were on hand about the comet's position in space to plot its future course with some accuracy, and it was obvious that Shoemaker-Levy 9 was going to get very close to Jupiter once again in July 1994.

In actual numbers, the comet seemed headed for an approach within 45,000 kilometers of Jupiter's center—and the planet has a radius of 75,000 kilometers. Excitement spread worldwide among astronomers. Never before had a collision between celestial bodies been known about before it happened, and this one came with six months' warning to get ready. Spirits collapsed when calculations showed that the 20 or so fragments making up SL-9, as the comet came to be called for short, would all make impact on the planet's far side, from Earth's perspective. Then it was realized that Galileo, in the summer of 1994, would be positioned just right relative to Jupiter to observe the crash sites, being the only witness in the universe.

The Second Asteroid: Ida and Its Little Moon

However, on the program before this completely unforeseen, irresistible opportunity—which was just what mission planners, still smarting over the failed antenna, needed—was the second asteroid flyby. Galileo had again ventured into the Asteroid Belt, approaching the asteroid no. 243, Ida. The critical approach navigation, which had gone so flawlessly with Gaspra, coincided this time with an acute crisis. The Mars Observer had fallen silent just before reaching its destination. Rescue efforts, which ultimately did not succeed, had taken up so much antenna time at the Deep Space Network that not all of the navigational photographs for Ida could be called up. The probability of getting sharp pictures of the asteroid declined somewhat.

Ida had also been the focus of extensive ground-based observations. With a diameter of about 30 kilometers, Ida figured to be about twice as big as Gaspra. It has a rotational period of 4.6 hours. And it looks like a three-axis ellipsoid with an axial ratio of 2.2 : 1.8 : 1. All of this could be inferred from light curves, the changes in the asteroid's brightness over time by a factor of two. Per rotation, Ida was showing two maxima and two minima.

Again, Galileo's handlers could hardly wait. Just a few days after the Ida flyby on August 28, 1993, they retrieved the first image fragments. Concerns that the camera platform had been improperly adjusted proved unnecessary. The pictures of Ida were sharp, revealing it

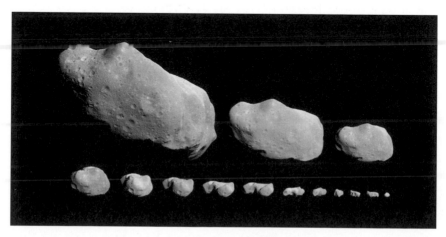

One "Ida day," captured in 14 images from a distance of up to 240,000 kilometers. The sequence encompasses approximately one complete rotation of the asteroid, which has a period of 4 hours and 39 minutes.

to be noticeably more irregular and jagged than Gaspra. On September 22, the first complete Ida photograph was produced—one of the first JPL products to be posted on the Internet, which at that time was known only to specialists. Within a matter of months it would become NASA's standard medium for publishing all of its new photography. News of the Galileo mission was made accessible to a large segment of the public in a way that would have been impossible just a year before.

At 52 kilometers in length, Ida turned out to be considerably larger than Gaspra. Also in contrast to the first asteroid, Ida showed clear evidence of numerous large impact craters. And one more thing—Ida had a moon! About 1.5 kilometers in diameter, it is much smaller than Ida, astonishingly round—and it raises a lot of questions. Whether it is "the first asteroid moon" or not is a matter of dispute.

Many times in the past there have been asteroid observations from Earth that suggest the presence of a moon. When these asteroids pass in front of a star, there is a second, very short interruption in the star's luminosity. Even the Hubble Deep Space Telescope has been called upon in the direct search for an asteroid moon, but with no success. So it was only among stellar occultation specialists that the existence of asteroid moons was accepted. For the great majority of astronomers, the picture of Ida's satellite offered the first clear proof. The tiny celestial body was christened Dactyl, putting yet another bit of Greek mythology into the heavens. According to the saga, the young Zeus was hidden in a cave on the mountain of Ida, so the craters Galileo discovered on the asteroid were named after famous caves on Earth. Other formations received the names of Galileo team members who had passed away. The craters on Dactyl were named after individual "dactyls"—magicians from Mount Ida on Crete who had the responsibility of watching over the children of Zeus and Rhea.

The essential questions, however, concerned the real Dactyl. Where had Ida's moon come from? Its spectrum and reflectivity (20

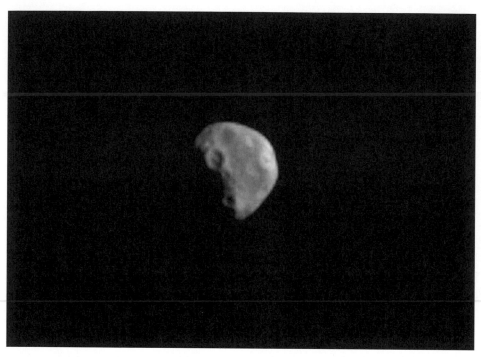

The most detailed picture of Ida's moon, Dactyl. It was taken on August 28, 1993, from about 3,900 kilometers, four minutes before Galileo's closest approach to Ida. Dactyl turns out to be shaped approximately like an egg, measuring 1.2 × 1.4 × 1.6 kilometers. More than a dozen craters with diameters of at least 80 meters are visible. The dimension at the terminator is 300 meters.

percent) corresponded almost exactly with Ida's, suggesting that a common origin is much more likely than some process whereby Ida would have captured Dactyl from somewhere else. Both are probably fragments from the same original body, the destruction of which is responsible for the entire Koronis family of asteroids.

Small differences in the spectra of Ida and Dactyl could be explained at least in part by "space weathering." In this interpretation, something that is not fully understood would have partially modified Ida, while completely transforming Dactyl. It is also possible that the two bodies come from different parts of their common predecessor, which themselves differed slightly in composition. That smaller bodies with moons could result in this way from the breakup of larger asteroids had been predicted in the 1970s. But there is a hitch. These same moons would then become targets for other asteroids and would quickly be destroyed. The likelihood, therefore, is that Dactyl is the rubble left over from a larger moon that has found its way back together after having been destroyed. Maybe such a process has been repeated many times, leaving Dactyl little more than a loose pile of rocks. This scenario would also explain how such a small object could have such a surprisingly round shape. Crater density, on the other hand, is just about the same on Dactyl and Ida, making it possible to "use" the satellite to learn even more about Ida itself. The series of photographs taken by Galileo's camera and NIMS was not quite adequate for identifying Dactyl's orbit exactly, but a number of possible orbits could be defined. That, in turn, allowed scientists to calculate Ida's density—which comes in as 2.6 plus or minus 0.5 grams per cubic centimeters, a medium value that is typical for chondrite-class meteorites.

The last pictures of Ida reached Earth in mid-1994, showing a surface with a wealth of geological formations. There were grooves, boulders, all kinds of craters, chains of craters, and dark and light spots. The number of craters implied the surprisingly advanced age of about a billion years. And craters over 1 kilometer in diameter are five times more common on Ida than on Gaspra. Ida, in contrast to Gaspra, has obviously not undergone any major "resets" of the surface, in which all the old craters are erased. This, however, does not necessarily mean that Ida had never been involved in any big collisions. A body the size of Ida—a proud 60 × 25 × 19 kilometers according to the latest readings—reacts much differently to any major impact than a small fry like Gaspra. Still, the mechanics of such processes remain very different from the results of collisions on large planets or moons. In the irregular gravitational field of an asteroid, the material thrown up from the surface when the crater is formed describes complicated ribbon patterns when coming back down.

Front-Row Seats for the Comet Crash

An altogether different collision occupied the Galileo planning team through the first half of 1994. How should the spacecraft be programmed to capture as fully as possible the series of crashes of the fragmented SL-9 comet about to occur on Jupiter? No one really knew how to predict how bright the explosions would be, or how long they might last. It was not even known exactly when the collision would happen, even though the comet's course was being tracked enthusiastically.

The problem was not recording a lot of data all at once. It was more deciding when it would be possible to retrieve the data to Earth. There was not much time for the slow process of data transmission. By the beginning of 1995, JPL would have to turn its attention exclusively to preparations for Galileo's arrival at Jupiter. The final decision was to document *all* of the impacts in one form or another, using the camera for some collisions and the other instruments for others. Real-time transmission would be possible for only one instrument, PPR, Galileo's onboard "light meter," which would need just a few bits to communicate simple light curves of possible lightning bolts on Jupiter.

And there was lightning in abundance! PPR immediately registered the impact of fragment H, the first part of the comet to crash into the planet—it raised Jupiter's total brightness by two percent inside of two seconds, after which brightness values returned to normal over the next 25 seconds. When the large fragment L struck, Jupiter's brightness increased by four percent.

Galileo's camera got its best shots of the great comet crash as "W," the last of the fragments, went down. A new picture was taken every 2$\frac{1}{3}$ seconds. There was nothing to be seen but the brief flashes on Jupiter's nighttime side, which is not visible from Earth. But even the exact knowledge of the timing of the flash helped scientists understand the complex processes at work when the fragment penetrated the Jovian atmosphere.

Just one month after the initial crash, Galileo's first images of the events began to arrive. Streaks made by the impact lightning were clearly visible, and the collision of fragment W was captured in a perfect series of individual flashes. Later on came other data, including readings from the UVS in the ultraviolet and NIMS in the infrared. The latter splendidly documented the way the fireball, which started off 8000° Celsius hotter than the surface of the Sun, quickly expanded and cooled. A series of major conferences throughout 1995 were devoted to analyses of countless observations made from Earth, from space telescopes, and from Galileo, all striving to form a conclusive overall picture of the singular event. What was observed to be happening precisely when and in which wavelength, and what did it all mean?

Not only Galileo, but also the Hubble Space Telescope and a variety of infrared telescopes had captured the impacts in astonishing detail. With each one, Hubble saw a mushroom cloud rise over Jupiter's horizon, quickly flatten into a pancake shape, and then drift down onto the upper layer of Jupiter's atmosphere. Infrared telescopes picked up lightning around the edge of the planet so intense it seemed almost excessive. Lightning would strike for minutes at a time, continuing to glow for hours afterward. The job now for astronomers was to put all of this into some kind of meaningful order—and it was the Galileo data that provided the key they were searching for.

Endless discussions later, most people who analyzed the observations agreed that the lightning seen by Galileo's camera and the PPR was from the bolide phases of the collisions, referring to the light created by the extreme speed (60 kilometers per second) of the comet fragment hurtling through Jupiter's atmosphere. From within the canal forged by the penetration of the fragment a fireball would gush upward in a final explosion—which turned out to occur at a temperature ideally suited for NIMS. Fifty seconds after impact, the fireball was big enough for observers on Earth to see it over Jupiter's horizon. At this point, Hubble's optical telescopes could see it too, and it had become just as perceptible for the infrared cameras of telescopes on Earth. For the infrared detectors, the ongoing effects of the collisions did not become really bright for another five minutes after that. By now, the images no longer had anything to do with the actual impacts themselves, but with how the "pancake" was settling back onto the atmosphere.

Without the Galileo observations, it probably would not have been possible to recognize all the connections among the various types of data so clearly. Nor would it have been possible to pinpoint the time of impact to the second. And now also the moment everyone had really been waiting for would not be long in coming. Galileo would soon arrive at Jupiter and its moons, and its true mission could finally get under way.

Chapter 3

Arrival and the Atmospheric Probe

Four Hundred Years of Jupiter Research

Jupiter. Júpiter, Yupiter, Giove, Zeus, Mushtarie. These are among the many names that have been given to this radiant celestial body in the night sky—which turned out to be just one of many. But how did research on the gas giant first get started? And what was at stake now, with the arrival of the first emissary from Earth?

"The giant planet had been regarded by mankind for millennia, but always as a mere point of light in the sky," says Thomas Hockey, a professor of astronomy at the University of Northern Iowa, who wrote his dissertation about Jupiter and published the book *Magnificent Planet—Observing Jupiter before the Photographic Age.* "The history of physically observing the planet began, abruptly, on January 7, 1610, when Galileo's first surveillance of Jupiter through a telescope revealed its planetary disk, thereby forever setting it apart from the stars. Not that Galileo saw anything on the planet's face. His telescope was not that good, and, besides, Galileo was much more interested in the three *satellites* he spotted orbiting Jupiter. A fourth turned up a few nights later."

Anyone with a simple pair of binoculars can reproduce the discovery today. But what sparked Galileo's interest in these points of light? "The satellites engrossed Galileo for three reasons," explains Hockey. "First, this confirmation of bodies revolving around a center not that of Earth supported the controversial Copernican, heliocentric theory. This was the theory to which Galileo adhered. Second, they aided his quest for patrons. Galileo promptly named the 'new planets' collectively for the Grand Duke of Tuscany, no doubt in the hope of reciprocal largesse. The 'Medicean Stars' never caught on, however. Eventually they became the Galilean Satellites. No fifth satellite of Jupiter was to be discovered until 1892."

But who caught the first glimpse of something actually on the surface of Jupiter? "A telescope maker, Francesco Fontana, Galileo's protegé Evangelista Torricelli, and the Jesuit theologian Niccolo Zucchi all have a claim on being the first to have observed Jupiter's characteristic dark bands in the 1630s, depending on which source is credited." According to Hockey, "the first astronomers to study the planet carefully, though, were definitely two professors at the University of Bologna: Francesco Grimaldi and Giambattista Riccioli. By the mid-1640s, Riccioli had watched the *shadows* cast on Jupiter by the Galilean Satellites, as they revolved overhead."

Of all seventeenth-century astronomers, however, Giovanni Cassini is the one with the strongest connection to Jupiter. He also came from

Bologna and later lived in Paris. It may have been the British Robert Hooke who first observed a spot on Jupiter, but Cassini was the one who used the movement of such spots across the planet to derive Jupiter's rotational period. Cassini also described a "permanent spot," which may be a reference to the initial discovery of the Great Red Spot. Otherwise, renderings of structures on Jupiter are rare, as Hockey laments. "Early observers spent little time describing the morphology of planetary features," because all they cared about were the numbers, specifically, the rotation period. But there is a 1711 painting of Jupiter from Bologna that shows a definite red spot in the southern hemisphere of the planet. "Who but Cassini (or his associates)," Hockey speculates, "might have motivated the painter to depict such an unusual feature?"

Jupiter became truly interesting in 1676 when it was used by the Danish astronomer and assistant at the Paris Observatory, Ole Rømer, to correctly determine the speed of light for the first time. The Galilean moons were always "late" whenever Jupiter was at its farthest point from Earth. By assuming that light does not propagate infinitely quickly, traveling instead at a certain speed, Rømer was able to make predictions of the moons' orbits work again—a milestone in the history of physics. "Once it was shown that the Galileans obeyed Kepler's Laws, Sir Isaac Newton's theory of universal gravitation received a major boost." Eighteenth-century astronomers, as Hockey points out, were relatively indifferent to Jupiter and the outer planets. Stellar astronomy, which served navigational needs, was more compelling. The exception to the rule was Sir William Herschel, an English immigrant from Hanover, Germany. He studied everything in the sky, so to speak, including the planets—and he made a fundamental discovery about Jupiter's moons. "It was Herschel," Hockey reports, "who correctly interpreted variations in the brightnesses of the Galilean Satellites." At different times we see the sides of the moons illuminated differently. "Correlating these variations with the satellites' orbital motions demonstrated that they, like Earth's Moon, rotate with a period equal to that of their revolution." They are, in other words, locked in "bound" rotations, always with the same side turned toward Jupiter.

Herschel also tried to improve Cassini's measurements of Jupiter's rotation, but he was frustrated by the transience of its structural features. And he, like Cassini, failed to consider the fact that the structures rotated at different speeds, depending on their distance from the equator. That condition is simply so different from what we experience here on Earth that it was completely outside the generalized image of the Solar System at that time. "Planets were expected to be, in the main, rigid bodies," Hockey says. "To Herschel we owe the model of a solid Jupiter with clouds blown across its globe by terrific winds in a murky atmosphere." No one, not even Herschel, had any doubts that eventually at some point under the clouds, they would find a definite surface. One school held that the brightest structures, the zones between the dark bands and the bright spots, *were* the surface. Others, like Herschel, thought it more probable that the dark bands and dark spots represented holes in the cloud cover, through which the surface could be seen. These scenarios were not as wide of the mark as it may seem. In the absence of evidence gained from modern infrared techniques, either of them, in principle, could be correct. There was an air of great confidence around Herschel's time that someday a hole would be found through which astronomers could get a look at Jupiter's solid surface. Once that happened, it would finally become possible to determine its rotation period exactly.

A number of well-to-do amateur astronomers from Germany stepped forward to make a critical contribution at this point in the history of Jupiter research. Among them was the pharmacist Samuel Schwabe of Dessau, best remembered for his discovery of the cyclical nature of Sun spots. More important specifically for Jupiter research was Johann Schröter, who studied the planet from his observatory in

Lilienthal from 1781 to 1813. Schröter was the first to speculate that the Jovian atmosphere might extend up to extreme altitudes. But even he chased the phantom of a "true" rotation period. Selenographers Johann Mädler and Wilhelm Beer introduced a bit more precision into the exploration of Jovian structures by using a micrometer to measure them. These two Berliners were probably the first to describe the "barges," a number of dark spots that are prominent features in Jupiter's northern hemisphere.

Friedrich Bessel, a mathematician and astronomer and founding director of the Königsberg Observatory, made no new discoveries about Jupiter's supposed surface. But he refined the calculation of Jupiter's mass and, therefore, of its density. The result—little more than 1 gram per cubic centimeter—drove home the point that Jupiter could not be like Earth, which consists primarily of rock and metal. It became obvious, if such a low mean density were to be explained, that hydrogen and helium had to be present in enormous amounts. That idea, in turn, helped clarify a feature that had been noted a full century earlier—that Jupiter has the shape of a flattened sphere. A rapidly rotating fluid body would automatically generate a bulge at its equator.

In the 1840s, a father and son team from the U.S., William and George Bond, made a series of drawings of Jupiter. Their pictures, as well as some by Franz Gruithuisen, clearly show the Great Red Spot nested in its characteristic hollow among the clouds. So they clearly knew of it. George Bond would later attempt to measure Jupiter's brightness, and he was the first to advance the thesis that more light came from Jupiter than it received from the Sun. The claim, in fact, is untrue, but since the 1960s we have known that Jupiter radiates very strongly in the infrared and that it has a much hotter interior than any body in the inner Solar System. A technique for measuring planetary brightness was finally perfected by Johann Zöllner at about the same time.

Jupiter's visible structures had always seemed sporadic and unpredictable to observers, perhaps because the planet had never been systematically investigated. This task would now be taken up by a group of British amateur astronomers. An early member of the group was William Lassell, a wealthy brewer. Toward mid-century, he discovered some white spots that appeared regularly in Jupiter's southern temperate zone. Such spots still exist today. They are visible on many pictures of the comet Shoemaker-Levy 9 crashing into the planet, very near the impact sites. A co-discoverer of the white spots was the rural clergyman William Dawes. He also described other large structures on Jupiter, and he was the first to document a major kind of turbulence that causes rapid changes in the brightness of one of Jupiter's bands. Astronomers by now were getting a glimpse of Jupiter's dynamic nature.

The question became, what could be driving all the activity? James Nasmyth was an engineer by training, but he was bold enough to publish an article on the nature of Jupiter's atmosphere in 1853, in the very midst of heated disputes among the professionals over planetary theory. His idea was that Jupiter had retained much of its heat from the time it was formed. Nasmyth proposed that the source of Jupiter's energy was not the Sun, but the ongoing process of cooling from an extremely hot initial condition. And in this he was absolutely right, although he also continued to believe that, far beneath the turbulent cloud formations, there would be the hot surface of a terrestrial planet. Nevertheless, Nasmyth's ideas were taken over by the popular astronomy writer Richard Proctor, and they helped push Jupiter specialists one more step away from the notion that it is an Earth-like planet.

The next important event was the "rediscovery" of the Great Red Spot in the 1870s. It had now become so conspicuous that it was impossible to overlook—likewise the curious fact that a large white spot was moving alongside it at a different speed. Finally, the differential rotation of the planet became established, and no one after 1880 tried to determine "the" rotation period.

Doubts were also being raised as to whether the giant planet had any kind of solid component in its interior. Parallel advances being made in solar research gave astronomers a way to distance themselves from the exclusive model of Earth-like planets. The Sun seemed to be a body similar to Jupiter. There were even attempts at the end of the nineteenth century to discover a physical link between the two giants. Observers looked for patterns in the long-term behavior of Jupiter's spots and their possible relation to the cycle of spots on the Sun. Speculation of this sort continued on through the 1960s. The Great Red Spot had finally drawn attention to Jupiter, and it was now being observed almost continually. "Still, at first, the modern astrophysics of the twentieth century seemed to have little to offer to the study of the Solar System beyond the Sun," Thomas Hockey states. "And professional astronomers were off in search of faraway galaxies."

Amateur astronomers had also maintained their interest in Jupiter, and they were rewarded in 1939 by the appearance of the White Oval, a major atmospheric event to be outdone only by the Great Red Spot. British observers meanwhile had been watching the planet for 50 years. Their data were used to justify claims that wind speeds in the various bands had remained astonishingly constant. All this was based on purely visual observations through a telescope, timing the progression of individual cloud phenomena across the surface—just as in Cassini's time. The first photographs of Jupiter had already been made around 1870, but systematic photographic observation became possible only in the twentieth century, at Lick Observatory in California and two observatories in Arizona. One finding made by this program, for example, was that the Great Red Spot and other ovals rotate counterclockwise. That identified them for certain as storms with no necessary connection to any solid surface. Despite such contributions, planetary astronomy did not really take off. This changed only in the 1940s, with the appearance of radio astronomy, the computer, and rockets. Around 1955, a periodic variation of Jupiter's natural radio signals finally betrayed the true rotation period of the gigantic interior planet. From this time on, it has been this inner rotation, and not transient atmospheric phenomena, according to which Jupiter's coordinates are defined.

Computers made it possible to model the complicated chemical reactions that take place at the various levels of Jupiter's atmosphere, assuming an initial composition like the Sun's. These models also predicted what kind of cloud layers there should be at which altitudes. With this knowledge, Jupiter became three-dimensional.

Rockets, finally, would free astronomers from the constraints of ground-based telescopes. Jupiter could now be explored at wavelengths that cannot be detected on Earth because they are absorbed by the atmosphere. More important yet, however, rockets gave scientists access to the planets themselves. Preparing for such missions into the unknown drastically increased the value of systematic observation from the ground. Hosts of amateurs and even a few professionals work, so to speak, as volunteers on the Galileo project, continuously monitoring the condition of the Jovian atmosphere. At the end of 1995, the urgent task was to understand conditions at the spot where the atmospheric probe would enter it. And in 1996, mission planners were grateful for tips on where in the atmosphere—given the meager rate of data transmission—the most exciting photographs could be taken.

Finally There: The Task of the Atmospheric Probe

At the end of January 1995, it was necessary to stop transmitting data from the great comet crash. Galileo and its atmospheric probe would be reaching Jupiter in just ten months. During the entire month of February, the Galileo orbiter would be fundamentally reprogrammed. New flight software had to be installed in place of the old, something that had never been done during a mission. The step was unavoidable. Only with the new "Phase 1" software package, in a certain sense Galileo's operating system, could the most critical tasks of the entire mission be accomplished: the release of the probe, the retrieval of its data, and the entry into a Jupiter orbit. And all without the high-gain antenna! Trading out the software went surprisingly smoothly. For just less than a year, it would control Galileo's destiny, until it too would be exchanged for the new "Phase 2" system, developed specifically for operations in orbit.

July 1995. Now things got serious. On the seventh, the probe was scheduled to be activated, and then released on the thirteenth. On July 27, Galileo's main engine, which until now had always been blocked by the probe, would be fired for the first time. It would be used to alter the course of the orbiter, which otherwise would follow the probe down into the clouds. The first task involved setting a timer in the probe so that the onboard electronics would "wake up" six hours prior to arrival. Once the timer was set, the connecting cable would be cut, Galileo having been turned to set the probe in the correct orientation relative to Jupiter. Finally, the clamps that for six years had been holding on to the probe were blown off.

There remained 83 million kilometers to go, but Galileo seemed to be perfectly on track. Another anxiously awaited event would be the firing of the main engine, which, like the thrusters, came from Germany. Just two years ago, an alarm had sounded for the Galileo team. The Mars Observer, just short of its destination, disappeared without a trace, and the investigation revealed that something had most probably gone wrong with its main engine. Could the same happen to Galileo?

Galileo was in fact not generally believed to be in danger, but contingency plans in case of engine failure were developed anyway. To be on the safe side, a two-second "wake-up" burn was conducted on July 24, before the main engine would have to burn for five minutes at a stretch to accomplish its major maneuver. Both burns went flawlessly. The change in Galileo's speed, down to 61 meters per second, was just under the target value (61.8 meters per second). Only one more minor course correction was necessary.

Then there was a problem. A vent had been jammed since the first time the main engine had been fired. A component of the helium pressure system, the vent's job was to keep fuel vapors from drifting in the wrong direction. Neither future course corrections nor the mission as a whole was felt to be threatened. But there were no experts to consult on how to eliminate the problem. The MBB engineers who had developed the drive system 20 years ago were mostly either retired or deceased.

Meanwhile, the Galileo orbiter had flown right into the middle of a genuine interplanetary dust storm. As many as 20,000 dust particles—fortunately tiny and harmless—were pounding against the detector, which had otherwise been registering one particle every three days. The storm dragged on into the fall, and expectations were that the count would not return to the previous low levels until the spacecraft had passed into the Jovian system. Where was all the dust coming from? Possible sources were not only the volcanic ash from Io and particles from Jupiter's rings, but also the dust brought along by P/Shoemaker-Levy when the comet crashed into the planet. While the fragments of the comet's core had disappeared in spectacular explosions, leaving behind a blackish residue, the dust had simply vanished. Had it turned into a reserve supply for Jupiter's exotic dust streams?

Then came another serious incident that once again brought the Galileo mission to the edge of failure. On October 11, the camera was taking three scheduled photographs of the entire planet, which would soon become too large to fit in its field of view. But when a command was issued to the tape recorder to rewind to the beginning of the data, it simply spun its wheels. Following a long break in communications, mortified flight controllers discovered that the pressure roller on the tape recorder was still spinning. For 15 hours it had worn against the same spot—somehow without breaking the tape.

It soon became apparent what had happened. The glue with which the data tape was attached to the roll had come loose and was now causing the tape advance mechanism to stick. The next week was filled with the direst imaginings. Would they have to do without the tape recorder? What kind of results could they expect in that case? Perhaps 150 to 300 extremely compromised images. The first hint of relief came on October 20. In an initial test, the tape moved forward with no difficulty at a lower speed, and it could be read. A new rule was issued: until the valuable data from the atmospheric probe had been recorded completely, if possible repeatedly, and read through, the recorder would be used only in slow mode. Taking additional photographs, even during the close encounter with Io during the Jupiter approach, was out of the question. Work at the higher data rate might become thinkable only in May 1996, after the new software had been installed.

The extremely sharp pictures of Io's surface, anticipated so eagerly for so long, were therefore out. Fortunately, plans for Galileo's atmospheric probe, which were impossible to postpone, were scarcely affected by the problems with the tape recorder. The data stream the probe would soon start transmitting was thin enough for the reduced tape speed to handle. Most of the data could also be written to the RAM of the onboard computer. There was suddenly some free space there, because the guidance software for the camera could be erased for the moment.

Arrival Day

December 7, 1995, drew steadily nearer, fortunately now with no further incidents. Then came "the longest day," as the Galileo team would name the rapid-fire chain of events just ahead. JPL, in its trusted fashion, scheduled the seven most critical hours for live broadcast on NASA television, including shots of the flight control center and ongoing commentaries, press conferences, and—perhaps—a closing celebration.

The show began at 1:00 P.M., California time, with a press conference announcing that the close flyby of Io had gone as planned. The closest approach, according to current course analysis, had taken place at 9:46 A.M., only about 6 seconds later than predicted. The minimum distance separating the orbiter from Io, about 890 kilometers, had turned out to be considerably less, however, than the targeted 937, leading to a larger than expected alteration in the spacecraft's trajectory. For Galileo it was pure gain, increasing fuel reserves by 90 kilograms, which would be available for future maneuvers.

More good news was that Galileo, despite the extreme levels of radiation in Jupiter's inner magnetosphere, had so far shown no sign of electronic failure. The Voyagers had experienced literally dozens of disturbances. A lesson had been learned, and Galileo's electronics had been made more robust. The JPL spokesperson was able to announce that even the tape recorder had worked flawlessly just an hour or so previously. All the ground stations were ready, and, as a precautionary measure, the stream of data from Galileo would be transmitted to JPL on no less than eight different channels.

There was still no radio contact at this point with the atmospheric probe. It had been "sleeping" since being separated from the orbiter and would awaken only when it was time to begin taking readings. Until then, the pressure could only intensify. It was not until midafternoon that the climax came. The probe must have been transmitting data for over an hour—and there was nothing from Galileo. Finally, at 3:11 P.M., came the eagerly awaited signal that Galileo had received the data from the probe. JPL was overcome with jubilation.

A second live shot in the control room brought more good news. It appeared that all the data were accounted for on the orbiter. Nineteen minutes remained to wait until another message of equally immense importance was expected. It would confirm that Galileo's 400-newton engine had successfully ignited for the 49-minute burn necessary to position the spacecraft into orbit around Jupiter.

All eyes were now on the so-called Doppler plot, a real-time graphic showing whether Galileo was diverging from a Keplerian orbit. If it did, flight controllers would know that the engine had ignited. During a long silence, the line stayed straight. And then, at 5:20 P.M., more cheers in the Mission Support Area, where someone had been the first to see the tiny initial deviation in the line. The burn came right on time to push Galileo gently into a Jupiter orbit, while by now the atmospheric probe had already disappeared into the depths of Jupiter's atmosphere. Contrary to widespread reports at the time, however, the probe was not crushed by the enormous pressures. The probe's housing was ventilated, equalizing the pressure inside and out.

The actual end was different. First, about 30 minutes after the probe had stopped transmitting data, the parachute would melt, and soon after that, at 660° Celsius and 280 bars of pressure, the probe's aluminum structure would follow suit. It would probably form a vapor cloud and accompany the sturdier remnants of the craft for some time. Not until temperatures reached 1,680° Celsius and pressure had mounted to 2,000 bars would the parts made of titanium melt, about nine hours after the probe first entered the atmosphere. One hour more and the atmospheric probe disintegrated into the individual atoms making it up, becoming part of the giant planet.

5:39 P.M.: from the Mission Support Area came the announcement that the engine had delivered nearly perfect thrust. The contribution from the German government, as was pointed out repeatedly, was a success. At 6:05 P.M., a spokesperson announced that Galileo would by about that time have become an artificial moon of Jupiter. The engine would continue burning another few minutes, however, to optimize Galileo's entrance into a complicated two-year trajectory from moon to moon. At 6:09 P.M., the engine was turned off. It was done. Galileo was in a Jupiter orbit on a nearly optimal course. Corrections would not cost much fuel.

Galileo's initial orbit described an extremely elongated ellipse, with a rotation period of 198 days, somewhat less than the planned 205 days because Galileo's Io flyby had been closer than expected. Instead of using fuel to alter the course, the first moon encounter—a Ganymede flyby—would simply be moved up a week. The mood was extremely positive, in short, and the final press conference of the evening evolved into a spontaneous celebration.

In just a few days, the first results from the atmospheric probe were supposed to become available. From December 10 to 13, technicians would be able to download part of the data being stored in the memory of the onboard computer, before Jupiter's position in Earth's sky came too close to the Sun, disturbing communications for a few weeks. A press conference was scheduled for December 19 by Ames Research Center, the NASA center in charge of the atmospheric probe. The plan was to end the year with a few initial impressions.

It was not to be. The data were there and so were the scientists, who did indeed have news to report. At NASA, however, operations had been shut down. A budget crisis had left the federal government technically insolvent a few days earlier, and all "nonessential" employees had been put on mandatory leave. Federal employees are strictly forbidden to work voluntarily and without pay, and violating the ban can result in a hefty fine.

Some scientists had shown up at Ames, those from nongovernment institutions and even from Germany, and they could not believe their eyes. Here NASA could fly a spacecraft to Jupiter, but it couldn't afford a press conference announcing the news! Angry scientists proposed convening the press conference in front of NASA's closed doors, but higher-ups managed to prevent them at the last minute. The U.S. managers of the atmospheric probe project, as federal employees, stood to lose their jobs.

This absurd news blackout, the first such event in the history of space exploration, went on and on. No sooner was the mandatory holiday over than the U.S. was hit by a severe snow storm. None of this stopped a tantalizing rumor from making the rounds, however, just as the year was coming to a close, that the probe had discovered a startling lack of water vapor in Jupiter's atmosphere. That would mean that on Jupiter there was not more oxygen than on the Sun, which was the universal expectation, but much less. Eventually it would become known from the same circle of amateur astronomers who had helped monitor Jupiter's cloud formations in the weeks prior to Galileo's arrival that the probe might have descended through an unusually cloudless region.

Still, not until January 22, 1996, could all the findings be presented, or at least those findings that scientists at this early point had already been able to distill out of the data. Specialists from all over the country and the world gathered to present an interim report at a disorganized but well-attended press conference. The media response was overwhelming. In practically no aspect did Jupiter's atmosphere seem to correspond to what theorists expected!

The Hour of Truth

The descent of the atmospheric probe was certainly the first climax of the Jupiter mission, for never before had an instrument manufactured by human hands come into such intimate contact with the gigantic gas planet. Never, for that matter, had human engineers faced such a challenge. The probe called for in the mission specifications would have to endure a drastic braking action caused by friction in Jupiter's atmosphere, then descend into the atmosphere, hanging by a parachute and taking readings—and all the while both temperature and pressure would be increasing to such extremes that it would be impossible to test the system completely in the laboratory.

The space engineers' mechanical creation is all the more deserving of recognition in retrospect. In a variety of ways, Jupiter had behaved very differently than had been assumed in the probe's design stage. Yet, all of the components stood the test—the robust capsule itself, as well as the instruments and the heat shields.

Galileo's atmospheric probe, with a total mass of 339 kilograms, was made up of two main components. First was the actual descent module, which would sink into the atmosphere and take readings with six different instruments. And second was the deceleration module, in which the instrument capsule was initially encased to keep it from burning up on its plunge into Jupiter's atmosphere at 171,000 kilometers per hour. This was the highest impact speed ever achieved by any system designed for spaceflight. Braking to 430 kilometers per hour would take only four minutes and would burn away most of the front heat shield. From an original 152 kilograms, only about 65 kilograms would be left at the end. For the probe to neither bounce off the atmosphere or, on the other hand, penetrate at too steep an angle and burn up, it had to enter at a precisely calculated angle.

Destination: the planet Jupiter as it was seen by Voyager 2 in 1979, here in a 1990 version reprocessed by the U.S. Geological Survey.

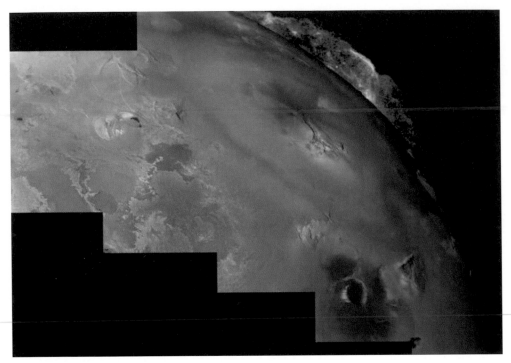

Eruption of the volcano Pele, as seen by Voyager 1 (in a new version first published in 1995). The plume rose to a height of 300 kilometers, spilling lava over an area the size of Alaska.

Liftoff! Atlantis takes off on Mission STS-34, with Galileo on board.

Galileo and its rocket stage being released. The trip will begin as soon as Atlantis retreats to a safe distance.

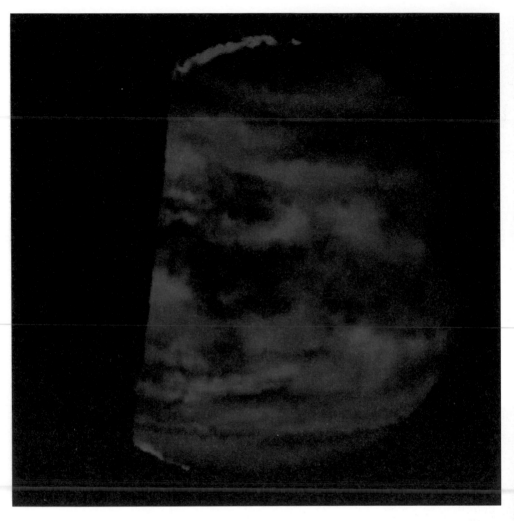

Intermediate-level clouds in the Venusian atmosphere, observed by Galileo's NIMS instrument at a wavelength of 2.3 microns. The red areas indicate thermal currents rising to an altitude of 50 to 55 kilometers through holes in the sulfuric acid clouds. The clouds that block this view in the visible spectrum lie about 10 to 15 kilometers higher.

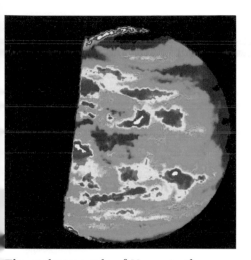

The nighttime side of Venus in the near-infrared at 2.3 microns, with the thermal currents rising from the depths partially blocked by intermediate-level cloud.

South America and Antarctica, as seen by Galileo on December 11, 1990, from a distance of 2.1 million kilometers. This is the first image in the famous film of Earth's rotation, which is made up of 500 real-color images covering 25 hours of weather and rotation-related phenomena.

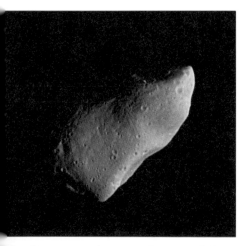

The first asteroid ever to be visited by a space probe: Gaspra from a distance of 5,300 kilometers.

Earth and Moon—individual image from the Galileo film during the second Earth flyby, showing the Moon in orbit around Earth. Since the reflective power (albedo) of the Moon is much less than Earth's, the Moon's brightness has been enhanced for this picture.

The asteroid Ida and its moon, Dactyl, taken from 10,500 kilometers. This picture was composed of images taken in the violet and infrared, which enhances the colors to an extent.

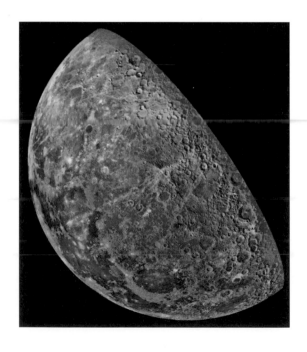

A false-color mosaic of the Moon composed of 53 photographs of the Earth-Moon system taken by Galileo in 1992. The color filters were selected to make variations in the composition of the Moon's surface visible. The old lunar highlands appear bright pink, and volcanic lava runs from blue through orange. The Sea of Tranquility appears dark blue here, because it is relatively rich in titanium. Mineral-rich areas, originating in relatively recent meteorite collisions, are bright blue, with the youngest craters shining bluest.

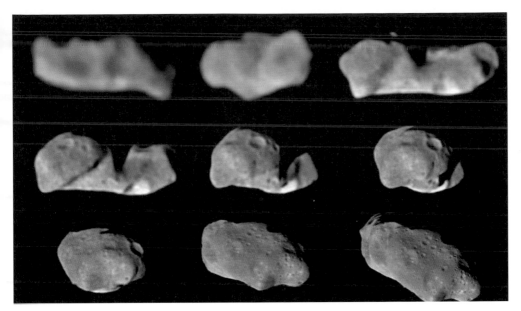

Ida's rotation in color. These images were generated in the violet and ultra violet spectra, but they were processed to approximate the way the asteroid would look to the human eye.

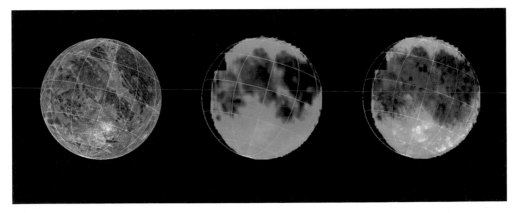

NIMS deciphers the surface of Jupiter's moon Ganymede, pictured at left in a Voyager photograph. The middle picture, because it was taken in the appropriate wavelength, shows the proportion of water (ice) among the materials making up Ganymede's surface (the brighter, the more water). On the right, mineral deposits show up red, and different sizes of ice crystals appear in the various blue tones.

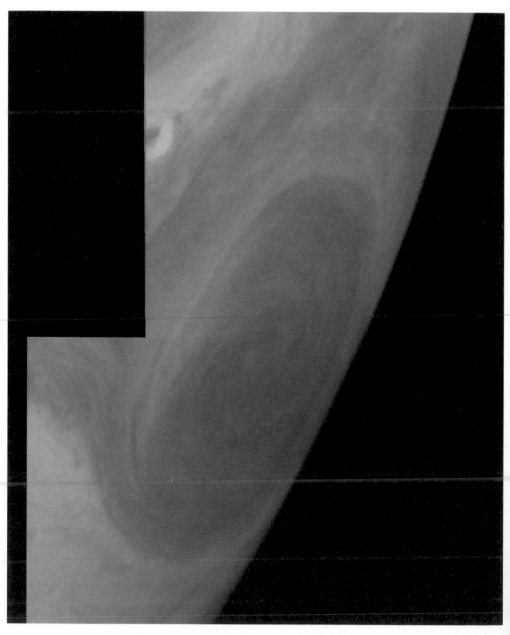

The Great Red Spot on the planet Jupiter: a long-lived storm system that measures more than twice the diameter of Earth from east to west and more than one Earth diameter from north to south (13,000 kilometers).

The Great Red Spot in false colors. By using the near-infrared filters on Galileo's camera and assigning them specific colors, image analysts are able to determine the altitudes of the various cloud layers. The deepest layers appear blue, and higher ones pink or, where the clouds are especially thick, white. Scientists have long known that the Great Red Spot is elevated (varying around 30 kilometers in altitude) and surrounded by particularly deep clouds.

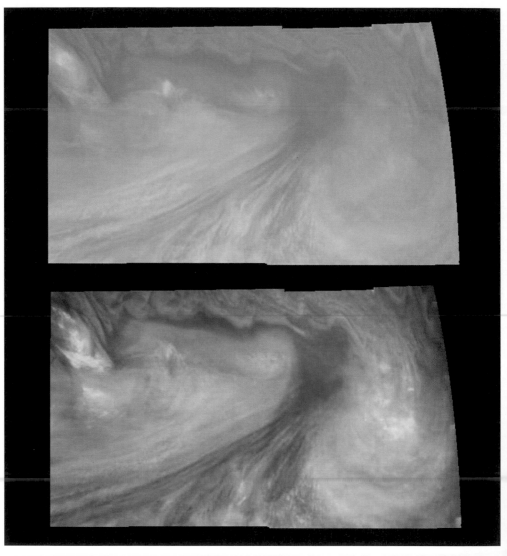

A hot spot in Jupiter's atmosphere in actual color and false colors. The top image is composed of the violet and near-infrared to correspond to what the human eye would see, and the bottom one was made using three different infrared wavelengths. In this bottom picture, color indicates altitude: bluish clouds are high and thin, red clouds are low, and white ones are high and thick. The dark blue hot spot turns out to be an opening in the cloud cover, albeit with a thin layer of vapor over it.

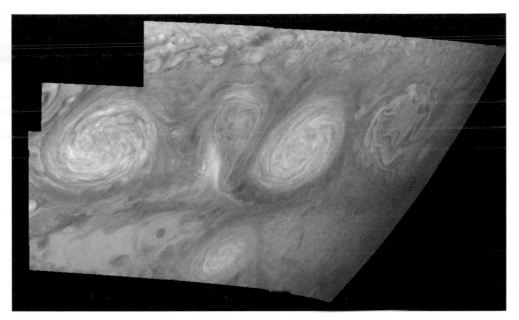

Jupiter's long-lived White Ovals in false colors, with a balloon-shaped storm pressed between them. The two obvious ovals are two of the three that have been known to exist since the 1930s and which, like the Great Red Spot, rotate counterclockwise (anticyclonic). The swirl in between rotates clockwise. The oval on the left has an east-west dimension of 9,000 kilometers.

Jupiter's northern hemisphere, between 10° and 50° latitude, again in false colors. The two connected storm centers in the upper half of the mosaic have a north-south dimension of 3,500 kilometers.

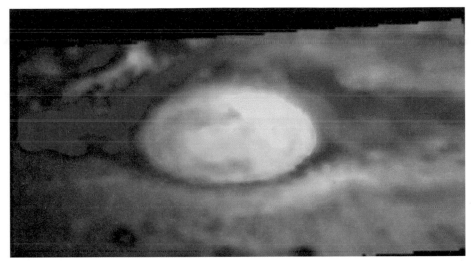

Jupiter's Great Red Spot in a completely different light. The colors red, green, and blue were each assigned to different infrared channels to emphasize the differences in altitude. The "green" color of the Great Red Spot means that these clouds are at a higher altitude than the clouds surrounding it.

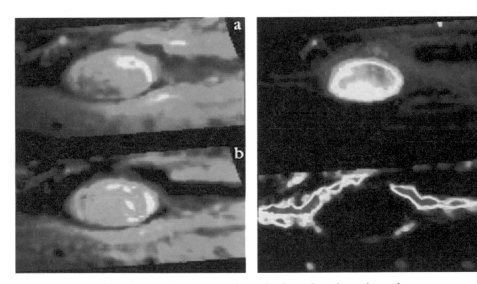

(a) The view in the deep red wavelengths, which is also about how these storm clouds would look to the human eye. (b) This shows the abundance of ammonia ice, with red indicating a lot, and blue little. (c) Cloud altitudes (red = highest, blue = lowest). (d) Jupiter's warm interior shimmers through the clouds. Red in this image denotes the thinnest clouds and blue the thickest.

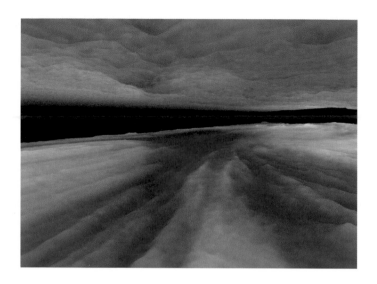

This is how Jupiter's clouds would look if a camera could be lowered deep into them near a hot spot. The light-colored clouds to the right of the hot spot could be examples of rising humid air and condensation.

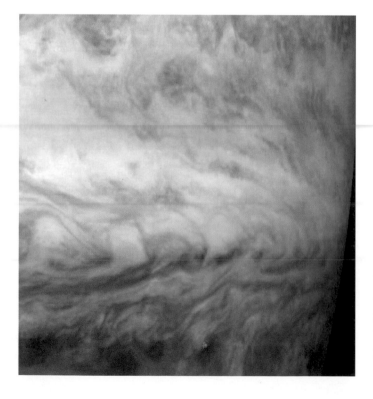

The transition between a cloud belt and a brighter zone near Jupiter's equator. This mosaic encompasses latitudes from 3° north to 13° south.

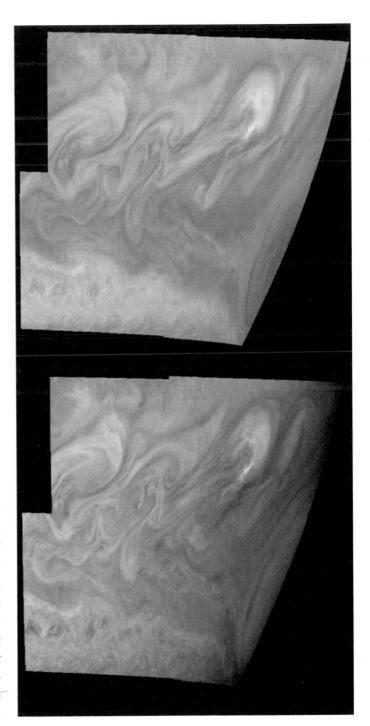

Turbulence to the west of—
and caused by—the Great
Red Spot. The storm center
deflects a westwardly moving jet stream to the north,
where it collides with one
going in the opposite direction. The swirls near the
Great Red Spot are lighter
colored, which gives scientists information about convection and cloud formation.

The boundary between belts and zones in the Jovian atmosphere, between latitudes 3° north and 13° south, on November 5, 1996, in quasi-actual colors.

Jupiter's cloud layers from a distance of 2.1 million kilometers. The top pictures on the left and right were taken at 1.61 and 2.73 microns, wavelengths in which the atmosphere appears relatively clearly. The clouds visible in these images are located at about 3 bars of pressure. The top picture in the middle, at 2.17 microns, shows only the very high clouds, because the hydrogen that makes up most of the atmosphere is opaque at this wavelength. (Only the Great Red Spot, the very highest clouds over the equator, and the vapor over the poles are still visible.) Yet another view is at the bottom left. At 3.01 microns, lower lying cloud layers can still barely be made out. Jupiter looks entirely different again at 4.99 microns (bottom middle). Visible here is heat being radiated from deeper cloud layers. Finally, at the bottom right, a false-color representation, in which red indicates thermal emissions from the depths, green the cool clouds of the troposphere, and blue the cold layers outside the troposphere.

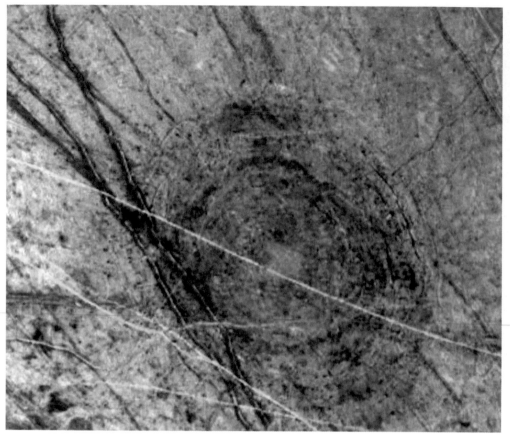

Tyre: an old impact crater, 140 kilometers across, showing the ongoing effects of the collision. First came the crater itself, then the red lines (from contaminated water ice), and finally the thin blue green lines.

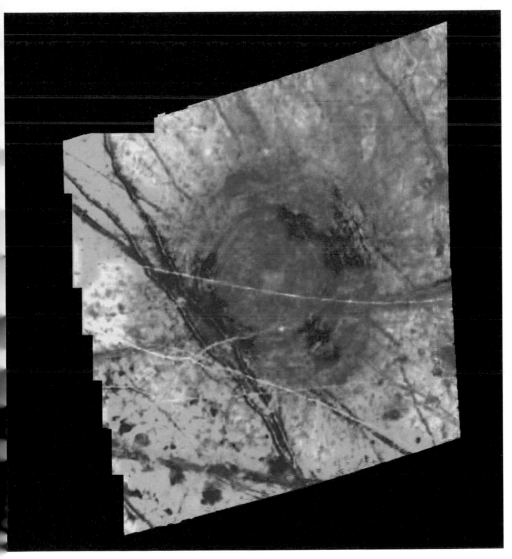

The composition of a planetary surface in detail, showing the Tyre region on the moon Europa, an old impact structure with a diameter of 140 kilometers. Blue indicates a high concentration of mineral salts—in compounds that are also found in the ground of California's Death Valley. Yellow-orange colors indicate large amounts of water ice.

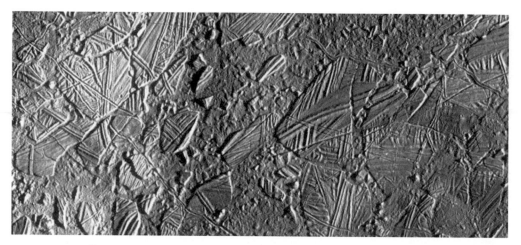

"Ice floes" on Europa and their color variations. Photographs from three successive flyovers of the Jovian moon were combined here to show not only the fascinating fine structure of the shifted ice blocks, but also the color nuances. The color data are much less resolved than the black-and-white pictures, but they make it possible to see small deposits left over from the formation of the crater Pwyll, which otherwise are partly covered up by ice of a more red color. These are not the original colors either, however, but come from mineral admixtures. Pure water ice would have a deep blue color and is indeed to be seen in other regions of Europa.

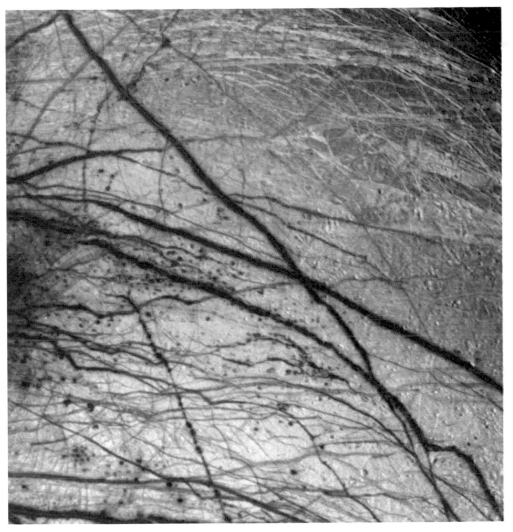

All ice is not alike; a clever selection of filters makes it possible to tease out subtle variations in Europa's ice crust. In some places, the ice has been contaminated by impurities, and in others, the physical characteristics of the material vary.

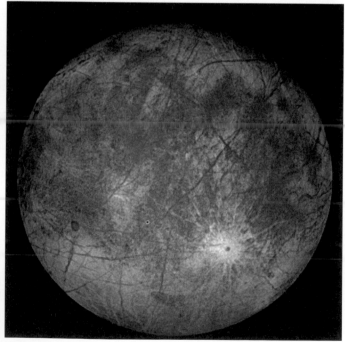

Europa more or less as it actually looks (top) and in enhanced colors. Mineral-rich material that has welled up from below appears dark brown, whereas clean ice that has been created in a variety of ways appears in various shades of blue. The bright object with a dark spot in the middle in the lower third of the images is the impact crater Pwyll, named for the Celtic god of the underworld.

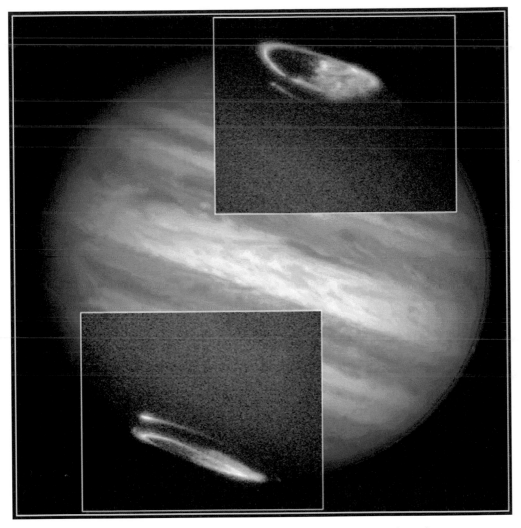

Jupiter's polar lights had never been seen so clearly, either from Earth or from a space probe. Much of the fine structure of the auroral ovals can be seen, as well as how the "footprints" of Io's flux tube leave obvious traces in their wake. (Photograph: John Clarke and NASA.)

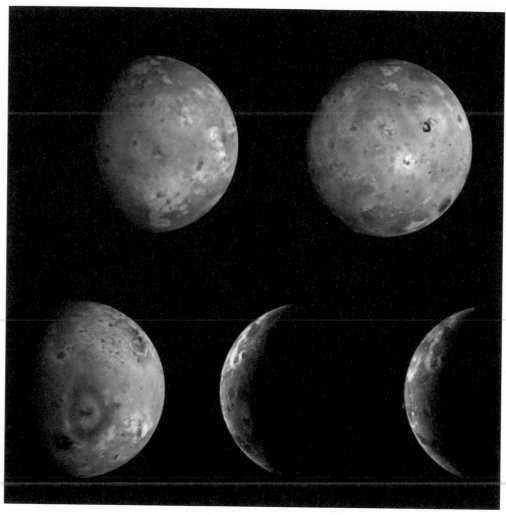

Io in various lights; the goal here is not just to register phases as we do of our own Moon, but also to examine the reflective characteristics of Io's surface from different angles.

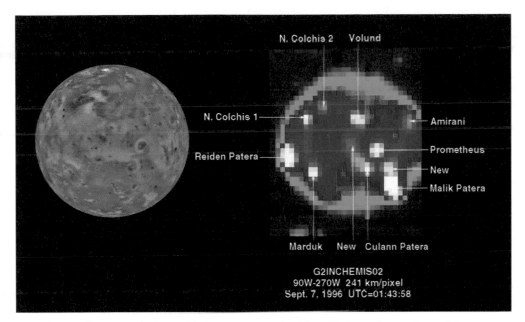

N. Colchis 2   Volund

N. Colchis 1 ——   ——— Amirani

Reiden Patera ——   ——— Prometheus
    ——— New
    ——— Malik Patera

Marduk   New   Culann Patera

G2INCHEMIS02
90W-270W 241 km/pixel
Sept. 7, 1996 UTC=01:43:58

Io in front of Jupiter. At least 11 hot spots can be identified on the surface of the moon.

Io in false colors to emphasize subtle variations in tone, which tell scientists about differences in the chemical composition of the lava. Red and orange indicate volcanic hot spots that presently have high temperatures, green signifies especially sulfur-rich regions, and blue-violet indicates sulfur dioxide frost.

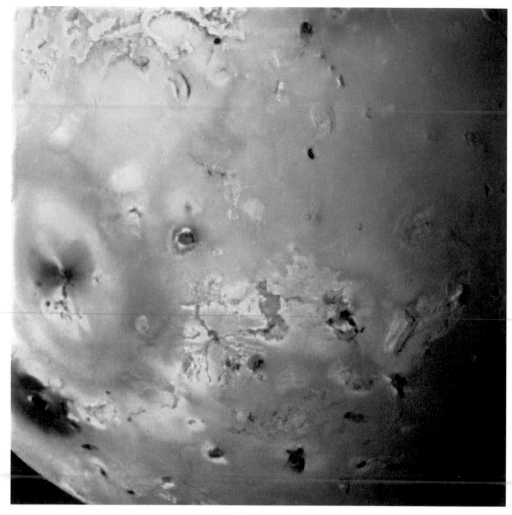

Deposits from the volcano Pele appear as a diffuse ring made of bright, reddish material in this combination of a color photograph and a sharper image in black and white. Pele was active at the time the pictures were taken in 1996, as was Marduk, likewise recognizable from the red deposits. If these red traces are not being constantly renewed, it takes only a few years for them to fade.

Io in front of Jupiter on September 7, 1996, from a distance of 487,000 kilometers—the highest resolution picture Galileo had taken of the volcanic moon up to that point. This is the side of Io which is turned permanently away from Jupiter. In the middle the active volcano Prometheus can be seen.

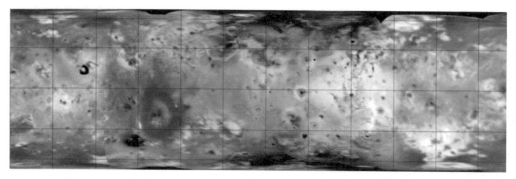

A flat map of Io, composed of false-color Galileo images from the summer of 1996. Sulfur dioxide frost appears white or gray in this picture, while yellow and brown indicate other sulfur compounds. Bright red or dark spots are fresh volcanic deposits.

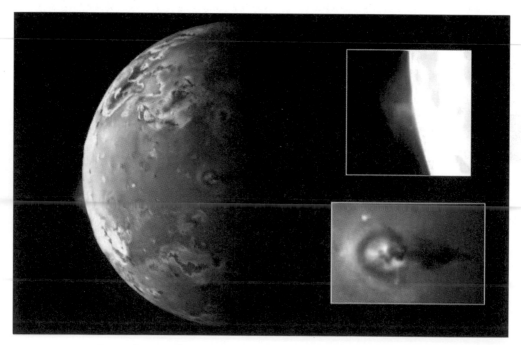

Two volcanic plumes on Io, one 140 kilometers high appearing on the edge of the moon over the caldera called Pillan Patera, the other over Prometheus near the terminator (the boundary between day and night). In late 1999, Galileo will fly directly over Pillan Patera, where the Voyagers had failed to detect any activity.

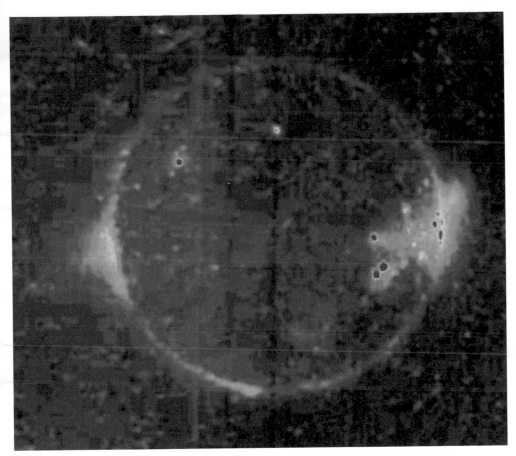

Jupiter's Great Red Spot in a completely different light. The colors red, green, and blue were each assigned to different infrared channels to emphasize the differences in altitude. The "green" color of the Great Red Spot means that these clouds are at a higher altitude than the clouds surrounding it.

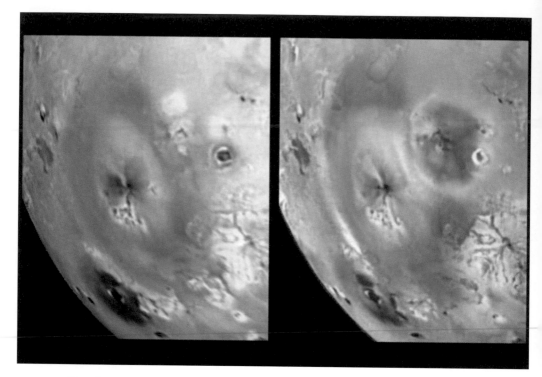

Big changes on Io since Galileo's arrival: a volcanic deposit the size of Arizona, 400 kilometers in diameter, was produced by the volcanic center of Pillan Patera between April and September 1997. In June 1997, Galileo and the Hubble Space Telescope both sighted a plume reaching 120 kilometers into the air, while infrared astronomers identified an especially high temperature hot spot. The gray color of the deposit is unusual. Most others are white, yellow, or red. This new one probably consists predominantly of silicates.

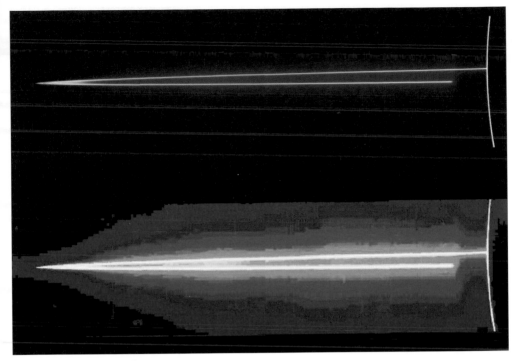

The main band of Jupiter's ring and its halo are visible in this false-color photograph. The halo, made up of tiny dust particles extending above and below the ring, is a curiosity seen nowhere else in the Solar System, and conventional celestial mechanics cannot explain it. Physicists expect that electromagnetic forces have a hand in the phenomenon.

This image mosaic from Callisto, covering an expanse of 1,400 kilometers, shows a pair of extremely old impact basins with a multiple ring structure. The crater Asgard dominates the region, with surrounding rings as much as 1,700 kilometers in diameter. A second formation, with rings up to 500 kilometers in diameter, appears to the north of Asgard and is partly obscured by the younger, brightly shining crater Burr. The ice released by relatively recent collisions is much brighter in appearance than the older surface of the moon, which is dark and reddish.

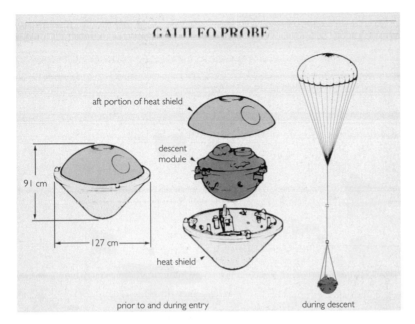

GALILEO PROBE

aft portion of heat shield

descent
module

91 cm

127 cm

heat shield

prior to and during entry

during descent

The atmospheric probe—inside the heat shield, and hanging by its parachute.

Some of the instruments had already starting working a few hours earlier—the charged-particle and lightning detectors, to name two examples. Even the drastic deceleration would be put to scientific use. The data would make it possible to calculate a density profile for the atmosphere. In a few minutes now, the probe would be dangling from its parachute, jettisoning what little was left of the heat shield.

About two minutes into the period of the most intense braking, starting 450 kilometers above "high zero" (defined as 1 bar of atmospheric pressure), the probe had descended to an altitude of only 43 kilometers, with all of the instruments functioning. Enough battery power remained for an hour and a quarter, during which seven fundamental scientific issues would be explored. German engineers were involved in three of the seven experiments and were project directors on two of them.

Galileo's atmospheric capsule: with no poetic name of its own, it was called simply the "Galileo probe."

What is the chemical composition of Jupiter's atmosphere? Exploring that question was the job of the Neutral Mass Spectrometer and the Helium Abundance Detector. The goal was to find out whether Jupiter's atmosphere continues to reflect the composition of the early solar nebula, or whether the gas giant had developed significantly since being formed.

What is the physical structure of the atmosphere? Equipped with sensors for density, pressure, and temperature, the atmospheric structure instrument would answer that question.

How high are Jupiter's clouds? What are they made of? The so-called nephelometer would shoot out laser beams and then register their reflections off droplets in the atmosphere.

Another instrument monitored radiation flux in the atmosphere. How much radiation moved down toward Jupiter at what altitude, and what were the currents going up?

A detector was keeping track of Jupiter's storms, which scientists hoped to document in both the flashes of lightning bolts and the

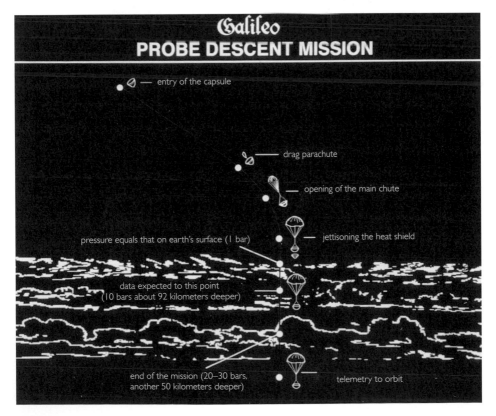

The probe's descent through Jupiter's clouds.

waves of electromagnetic energy they released. This instrument was combined with a charged-particle detector that was especially designed to explore the innermost region of Jupiter's magnetosphere.

The final question the probe was going to answer concerned wind speed at different altitudes. How characteristic were the major cloud movements that had been under observation for centuries for large-scale wind movements? The probe's radio signal was monitored as a way to evaluate the wind, once by the receiver on board the orbiter, and again by the 27 antenna dishes of the Very Large Array (VLA) radio telescope on Earth. The idea for this bold experiment came only long after Galileo was launched. Not even the largest antennas on Earth could pick up much of the extremely weak radio signal itself. The VLA data would be critical, however, once the probe's "broadcast" as a whole had finally been transmitted from the orbiter's memory banks. Using the two records together, analysts would be able to distinguish the signal from the noise otherwise picked up on Earth. And then it would be possible to determine wind speeds more precisely, because the two antennas picking up the signal, on the ground and on the orbiter, received it at different angles.

Analysis

The broadcast "live" from Jupiter's atmosphere had lasted just 58.6 minutes, covering an altitude range of 160 kilometers. It had begun 53 seconds late, with the initial readings starting at 0.42 bar, rather than 0.1 bar, of atmospheric pressure. The reason for the failure was quickly discovered, and a bit embarrassing. The two deceleration sensors on the probe responsible for starting the readings had mistakenly been wired to each other's ports on the orbiter's computer! There were two switches on board which were set to trip once a certain deceleration had been achieved—and the one sensitive to more drastic braking was programmed to react before the one keyed to a more moderate slowdown. Naturally, the opposite made sense. Only because of the tolerance for error designed into the computer programs did the instruments get started at all.

The end of data transmission came at a temperature of 150° Celsius and 24 bars of pressure, which led first to the batteries being rapidly exhausted and then to the complete failure of the transmitter. The inexorable destruction of the atmospheric probe had begun, while 200,000 kilometers above, Galileo continued on its course. The capsule had survived a surprisingly deep plunge into the atmosphere, far beyond the 10 bars of pressure called for in the initial specifications.

Analyzing the data, however, proved considerably more difficult than scientists might have hoped. A design flaw in the probe itself was probably responsible. The temperature variation inside the probe during the descent had gone far beyond the range for which the instruments had been made. All of them were affected, if to different degrees. In all likelihood, the temperature inside the probe was probably tied more strictly to outside temperatures than anticipated. That sent researchers back into their laboratories to sort out how to recalibrate the instruments for the new temperature range. For some of the devices, a few months is all it took, while other teams found themselves still at work two years later. Even the most spectacular announcements of the January 22 press conference came in for correction to some extent. Nevertheless, by May 1996 at the latest, most of the highlights were known:

• Helium is much more abundant than expected on the gas giant, at a level approximating the Sun's. In January, the opposite had been maintained.

• Carbon and sulfur levels had been predicted approximately, but very little oxygen was detected—which should have been present in the form of water vapor. This very early finding, while correct, could not be generalized.

- Unfortunately (and ground-based monitoring of the site of the probe's descent left no doubt on the matter), the readings were not typical. The probe had landed more or less in the Jovian equivalent of the Sahara desert, on the southern rim of a hot spot, and the "missing" water was present in abundance elsewhere. A substantial extra bottom layer of Jupiter's clouds had been predicted, which surprising early results had indicated was not there. But that need not apply to the whole planet.

- The bewildering finding that there seemed to be hardly any atmospheric particles and, especially, that the sharply defined layers that were thought to exist seemed not to—these puzzles, too, might merely represent local anomalies.

- Wind speed data confronted analysts with puzzles for a long time. At the end of 1996, it had seemed well established that, at 20 bars, Jupiter's winds would get even stronger. This seemed to be definitive evidence that the motion of Jupiter's clouds is driven by internal heat sources. By August 1997, analysis was revealing a different picture. Wind speed remained constant (at 170 meters per second) over the entire period the probe was taking readings, making the question of what drives the winds a mystery once again.

- On storm conditions, some suspect radio waves were picked up, but no optical signals. The radio waves (ranging from 3 to 15 kilohertz) came from a great distance and were not much like the radio emissions given off by our lightning on Earth. Lightning bolts on Jupiter are somewhat stronger, but also more rare, than they are on Earth.

- Another success, finally, was the exploration of the magnetosphere's innermost region, where charged particles exist in abundance, forming a previously unknown radiation belt.

Scientists wrestled with the data in the years to come, interpreting the results and determining whether they could be applied to all of Jupiter. The enormously arid region, probably a "hot spot," into which Galileo had released the probe in December 1995, was really anything but typical for the giant planet. These were the especially cloudless regions, making up just one percent of Jupiter's atmosphere, through which it is possible to see warmer, deeper layers. Still, would it be possible for such hot spots to be so thoroughly "dried out"? For some atmospheric scientists, such findings posed no particular problem. They could be the effect of powerful atmospheric downdrafts. According to this scenario, the atmosphere near Jupiter's equator is warmed by the Sun, causing the gas to rise. Clouds form and the water disappears by condensing into droplets (or ice crystals). The dry air then flows to the north and south to descend in true "desert" regions, circulating almost exactly like on Earth.

This idea is confirmed by the systematic observations of Jupiter conducted with Galileo's infrared NIMS camera, as well as by data from special ground-based telescopes. Humidity varies considerably from place to place in Jupiter's atmosphere—as indeed it does on Earth. The hot spot into which the probe descended (entirely by accident), in this sense, might be compared directly with the Sahara or with California's Death Valley. Only two to five percent of the planet is likely to be this dry, however, with humidity levels nearby as much as 100 times greater, perhaps rising high enough in some places for there to be "rain." So, the most fundamental result of the probe's kamikaze plunge into the Jovian atmosphere might be altogether banal: Jupiter is much more like Earth than we thought!

Progress Report: Where Did Jupiter Come From?

If anything, the data from Galileo's atmospheric probe have raised more questions than they have answered. One aspect understood now better than in 1995 is the physical environment into which the probe had dropped: The workings of such a hot spot have been simulated in detail in computer models. Waves of up-and-down winds that span great ranges in air pressure apparently can explain these surprisingly clear, dry areas near Jupiter's equator. "If you could ride in a balloon coming into one of the hot spots, you would experience a vertical drop of 100 kilometers," explains Jupiter atmosphere specialist Andy Ingersoll. The model shows that air moving west to east just north of Jupiter's equator is also moving dramatically up and down every few days. Water and ammonia condense into clouds in Jupiter's white equatorial plumes as the vapors rise. Then the wrung-out air drops, forming the clear patches—into one of which the probe happened to plunge by pure chance. After crossing those hotspots, the air rises again and returns to its normal cloudy state.

The computer simulation reveals that the probe's entry site is probably even more unusual than previously thought. Both the probe and the computer model show that the head winds on the southern rim of a hot spot get stronger and stronger with depth into the planet. But in the model, this trend is reversed on the northern side. These results underscore the importance of future multi-probe missions to Jupiter. The hot spots were known for a while, but their depth was a surprise. A better name for them may be bright spots, since the temperature at their visible depth is only about 0° Celsius—still relatively balmy, though, compared to the neighborhood of −130° Celsius at surrounding cloud tops. Each hot spot is about the size of North America and lasts for months. The hot spots alternate with larger cloudy plumes in a band near Jupiter's equator. In some ways the dry areas where wrung-out air masses are descending resemble subtropical deserts on Earth, but unlike our planet, Jupiter has no firm surface to stop the air's fall. All the hot spots combined make up less than one percent of Jupiter's global area, but understanding how they remain stable is important for the whole planet's dynamics.

Even if the probe entered a rare region of Jupiter's atmosphere, the relative proportions of its gases found there should be pretty typical of the planet as a whole—and those relative abundances have become rather perplexing as the complicated analysis of the probe data—that the scientists themselves describe as "agonizing"—went along. Apparently the gas giant contains two to three times more of the heavy noble gases Argon, Krypton and Xenon than one would expect had it formed solely from the solar nebula, i.e., the leftovers from the formation of the Sun. Jupiter also contains about three times more nitrogen than would be expected under the prevailing models of how the Solar System was formed. Where Jupiter now orbits, about five times as distant from the Sun as the Earth (5 AU), it is much too warm to have accumulated those gases in the quantities detected by the probe. Jupiter is believed to have formed from the solar nebula and from a collection of small bodies or icy planetesimals. Most planetesimals (to which also the comets belong) are thought to have formed somewhere between the orbits of Uranus and Neptune, 20 to 30 AU from the Sun.

Even at that distance, however, the initial temperature of these icy bodies would have been far too warm for them to trap the heavy noble gases and nitrogen in an icy form. An alternative scenario involves the delivery of these gases to Jupiter by small icy bodies that came from the Kuiper Belt, 40 AU from the Sun and beyond the orbit of the planet Neptune. But had the planetesimals somehow fallen from their orbits in the Kuiper Belt to Jupiter's current orbit, the noble gases and the nitrogen would have largely dissipated in the warmer temperatures before they arrived at Jupiter—thus neither theory accounts to what the probe found in 1995. Could it then be that Jupiter was initially much farther from the Sun and has moved into its present orbit only recently? It would have formed 40 or 50 degrees from the Sun in that scenario, at a chilling temperature of $-240°$ Celsius, only to move ten times closer to the Sun at a later stage. Orbital migrations of big planets by some degree are considered not too strange in modern views of the Solar System: It's either that, or the solar nebula was much, much cooler than present models estimate.

The Cruise Begins

Galileo's actual journey through the Jovian system was set to begin only after the adventures with the atmospheric probe had ended. To get started, it was necessary to get a third and final burn from the main engine, for 24 minutes on March 14, 1996. The precisely executed maneuver was more than a simple course correction in the orbit initially taken around Jupiter. Galileo's speed nearly doubled, and for the foreseeable future, it would come nowhere near the orbit of the moon Io. The point of this major maneuver was to shift the perijovium, the point in Galileo's orbit nearest to Jupiter, far enough out from the planet for the spacecraft to escape the intense radiation of the inner zone. The minimum distance separating Galileo from Jupiter jumped from 185,000 to 786,000 kilometers, creating more secure conditions for the grand two-year orbital tour to come. The orbiter's primary mission consisted of the three close flybys of the remaining Galilean moons, Ganymede, Callisto, and Europa, at distances between 250 and 1,600 kilometers, with five additional observations planned from distances of 3,000 to 80,000 kilometers. There would be panoramic views from unusual perspectives, and conditions would be optimized for monitoring volcanic activity on Io. The best closeups were anticipated to outdo the resolution of even the sharpest Voyager photographs by a number of orders of magnitude.

To put this in perspective, Galileo's advance over Voyager could be compared with the difference between looking down at a book on the street from the top of the Empire State Building and, with Galileo's help,

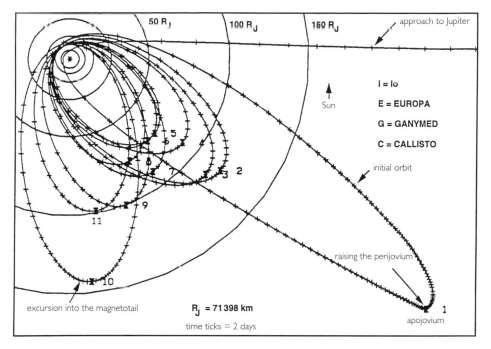

The grand tour of the moons—primary mission.

having the book at arm's length, reading. The increase in resolving power between the first real pioneer, Galileo Galilei, standing with this telescope observing the moons in 1610, and the spacecraft Galileo visiting them in 1996 and 1997, is measured by a factor of 100,000 to 1 million.

A new flyby of one of Jupiter's major moons would occur about once every two months and would also serve to change the orbit. In this way, Galileo would travel from one destination to the next, sometimes visiting the same moon again, sometimes moving on to a different one, at the lowest possible fuel cost. The revolution in the strongly elliptical orbit was always the same. Each time around, the closest approach to the respective destination lasted for a week. The tape recorder, which was now working flawlessly, would put pictures and other data on tape, and in the ensuing months between visits, the information would be slowly processed and transmitted through the low-gain antenna back to Earth. For the planetary scientist on Earth, in other words, the "Jupiter experience" was the reverse of what happened with the Voyagers: the results were always there when Galileo was *not* close to a moon. The only constant readings taking place now were the work of the dust counter and the field and particle detectors in the magnetosphere. The latter was examined in particular detail during a major "excursion" of Galileo's into the magnetic tail dragged by Jupiter in the solar wind, which lasted nearly the entire summer of 1997.

For the very last time, engineers had tried to open the high-gain antenna on March 25, 1996, just after the perijovium was raised. It was, as expected, futile. The shaking it had been given when the main engine fired had evidently not done the trick either. Now, the orbiter

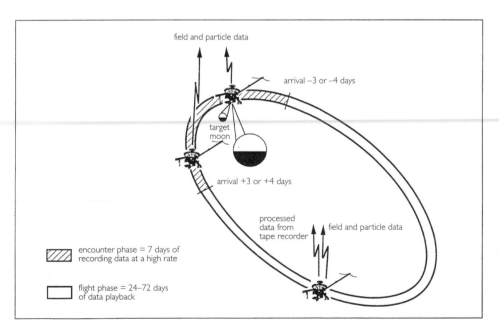

field and particle data

arrival −3 or −4 days

target moon

arrival +3 or +4 days

processed data from tape recorder

field and particle data

encounter phase = 7 days of recording data at a high rate

flight phase = 24–72 days of data playback

A typical Jupiter orbit for Galileo consists of one week with the instruments taking readings near a specific target moon, followed by one to three months of data transmission.

mission emergency plan, the product of four years' preparation and improvement, automatically went into effect. In May and June, most of the orbiter's software was exchanged. Brand-new, for instance, were clever algorithms capable of compacting the data, especially the pictures, with maximum efficiency. The benchmark, established as early as the end of 1991, was to achieve "an effective data rate" of 1,000 bits per second. The high-gain antenna, for the sake of comparison, would have transmitted 134,400 bits per second. If they made the benchmark, practically everything the recorder could put on tape (about 900 million bits) could be retrieved between moon visits. Technicians reprogrammed all of Galileo's computers. Antennas on Earth were optimally arranged. And, in addition to all that, Galileo would transmit data at a variable rate to conform with the availability of antennas. Still, the big boost would come from data compression.

The big test would come at the end of June 1996, when Galileo paid its visit to Ganymede. It was time for the tour to begin.

What follows is a systematic description of the biggest surprises and most important discoveries of the primary mission, in which the lessons and pictures for each "location" are summarized along with the stories of the respective visits. Since the first super-sharp pictures and fascinating data came from Galileo's initial moon encounter, we first turn to Ganymede.

Chapter 4

Looping from Moon to Moon

Ganymede: A Ball of Ice with Furrows and a Magnetic Field

Six months after becoming a part of the Jovian system, Galileo began a slingshot journey that took it from moon to moon. The first attraction came on June 17, 1996, when the orbiter flew within 835 kilometers of Ganymede, the largest planetary moon (at 5,270 kilometers in diameter) in the Solar System. It is larger than the planet Mercury, but its density is so low (1.9 grams per cubic centimeter) that it must contain large amounts of ice. The "geology" that emerged in the Voyager and

Galileo photographs consisted not of rock but of frozen water! The Voyager pictures had already revealed a long and turbulent history of the ice ball, including plate movements of the crust and the formation of mountains. Ganymede's outer shell seemed to be broken up into blocks that were then shifted dozens of kilometers laterally. Long parallel mountain ranges and valleys cut across other regions, as if a gigantic rake had been drawn across the surface, scratching "grooves," as experts called them, into the surface. It was probably in undergoing a general expansion that the surface broke up into the pieces that had drifted apart. That could have occurred, to name one possibility, when the relatively light ice separated from the heavy constituents of the moon's progenitor body.

This eventful period, encompassing several phases of mountain formation, would have lasted perhaps 1 billion years. Not much has happened in the last 3 billion years—or at least that is how things seemed from the Voyager pictures. A lot of thought had been devoted in the meantime to what sort of mechanisms could have led to the astonishingly ordered geological system. No theory really worked, however, to predict the groove pattern in quantitative terms. Finally, a critical step forward seemed to have been taken at Washington University in St. Louis. In the early years of the Solar System, the Sun was about 25 percent less luminous than it is today. In the Washington University study, scientists took into account not only this established fact of astrophysics, but also new knowledge about the dynamics of extremely cold water ice.

The theory, which now seemed to work quantitatively, treated Ganymede's crust as a fragile surface covering some kind of malleable substructure. With this model, scientists were able to calculate how the outermost crust became deformed when the moon expanded, while the pattern of ridges and valleys which was predicted recalled Ganymede's grooves. With new data on the temperature of the ice (colder than expected) and how it behaved under stress, the model predicted the emergence of the right size grooves—a few hundred meters' relief at rather evenly spaced distances averaging about 8 kilometers. Now, however, sharper images of the icy landscape had become available, and matters were looking more complex than the Voyager pictures had suggested.

Galileo's encounter with Ganymede in June 1996 was the first of as many as 25 planned visits to the Galilean moons, and it came off flawlessly. This was just a foretaste of the encounters to come, which would repeat every one to two months through the end of the extended four-year program. First came a long dull string of telemetry data, the fruit of Galileo's successful work during the critical weeks of its closest approach to Jupiter and to the moon. Navigational maneuvers had clearly been successful. Flight control was able to follow them in real time as well, because of the characteristic effect of the moon's gravity on the speed of the orbiter. The "pitch" of the radio signal underwent the same variation, as the Doppler effect either slightly compressed or elongated the radio waves whenever the spacecraft was accelerated or braked. Information beyond instrument status reports and course coordinates was not available during the actual week of the visit. The tape recorder was completely occupied storing all the different types of scientific data.

As soon as the visit was over, with Galileo once again tracing a broad ellipse through the outer reaches of the Jupiter system, scientists began the process of reading all the valuable data off the tape, often several times through, with the process lasting until just before the onset of the next encounter.

Being a Galileo scientist at this point of the mission, in other words, mainly meant waiting—no less for involved amateurs than the JPL pros. Data often arrived in unconnected bundles, and often it took weeks for complete images to develop. From Galileo's initial visit to

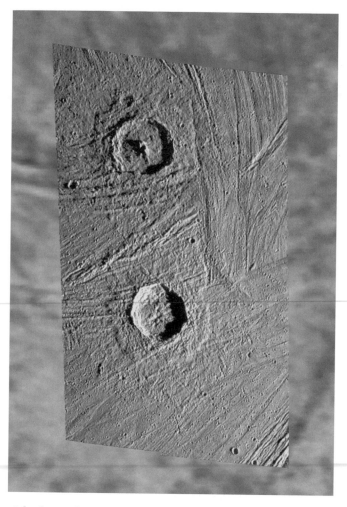

A high-resolution mosaic of the craters Gula (top), about 40 kilometers in diameter, and Achelous (bottom), about 35 kilometers in diameter, projected on a lower-resolution background of image data obtained in the late 1970s by the NASA Voyager spacecraft. The two Galileo frames were taken under low sun elevation in April 1997, during Galileo's seventh orbit, from a distance of about 17,500 kilometers from Ganymede. The resolution is about 180 meters per pixel; the smallest features that are still discernible are about 360 meters across. A characteristic feature of both craters is the "pedestal"—an outward-facing, relatively gently sloped scarp that terminates the continuous ejecta blanket. Similar features may be seen in ejecta blankets of Martian craters, suggesting impacts into a volatile ice-rich target material.

Ganymede, an astonishing 127 of the 129 photographs taken ultimately reached Earth. Frustration levels could rise even higher for
Jupiter fans being forced to wait for new pictures to be published on
the Internet, where the publication rate had dwindled to about one
picture per workday, which was likely to be highly processed and
months old. JPL swore its good intentions. There really was no way to
process the data any faster. There were exceptions every couple of
months, and then NASA always held a press conference in regard to
some particularly spectacular discovery. Completely new pictures
would be shown, which even researchers had hardly seen. Spontaneous interpretations were delivered on the spot, a ritual of the "instant science" that no one does better than JPL.

Developments had reached that point on July 10, 1996, following
Galileo's first week-long encounter with Ganymede. The world got its
first close look at the moon's icy landscapes, and it learned with a shock
that this otherwise dead-looking ball of ice had its own magnetic field.

The Magnetic Moon

Galileo's plasma wave system was the first to register something unusual. The device can be thought of as a large, high-amplification radio
antenna tuned to pick up all radio signals at once. Readings can be either represented in image form or broadcast over a loudspeaker.
Plasma wave researchers develop a sixth sense for these sounds. With
some training they can actually hear the physics behind the waves, and
their jargon likewise tends toward the acoustical. About 45 minutes of
"loud radio noise" during Galileo's closest encounter with Ganymede
is what signaled the magnetic phenomena to them. In contrast to the
case of Io (see pages 230–231), which would have plasma specialists
arguing for years over interpretations of the Galileo data, the judgment
on Ganymede was clear from the first two visits: this moon had a magnetic field, which it in all probability generated itself (just like Earth)

by means of a core dynamo of molten metal. Down to the details, the effects corresponded to comparable observations near magnetic planets. Typical signs are the loud "crack" heard upon entering and exiting the magnetosphere, the "whistler" waves, or magnetosphere radio emissions. Ganymede turned out to be the first moon in the Solar System to be a nonthermal radio source!

Since Ganymede is located completely inside Jupiter's magnetic sphere, forming, so to speak, a magnetosphere inside a magnetosphere, there were all kinds of unusual reciprocal effects to be noted. A total of four different times, Galileo's magnetometer also confirmed Ganymede's magnetic field directly. It is without question a dipole, like Earth, or a bar magnet, which argues for a source deep inside the moon. The rotation axes of the planet and its magnetic field are inclined at only ten degrees, which is nearly parallel. This fact also conforms to a dynamic of rising and falling currents in a metallic core. An earlier hypothesis, according to which the field could be generated by a salt-water ocean, has by now been largely discounted. The goings-on deep inside the moon had already been exposed by Galileo's readings of Ganymede's gravitational field. The core is clearly dense, from 5 to 8 grams per cubic centimeter, making it the most "centrally condensed" solid body in the universe. A model of the moon emerged in three parts. Ganymede has a partially liquid metal core, which seems to be the location of the electric dynamo. It has a silicate mantle, and an outer shell of ice. Without the ice, Ganymede would be almost exactly the size of Io and have nearly the same internal distribution of mass.

Ganymede's iron core spoke volumes about the moon's history. At some point in the past it had to have been heated to a temperature of at least 1,052° Celsius. The heat either released by the formation of a "protojovian" disk out of matter or created by an early abundance of ra-

dioactive material might have been just enough to have done it. Yet another possibility would be that Ganymede had once been in a different
orbit around Jupiter, subjecting it to the kind of tidal heating that Io
continues to undergo—and which has such remarkable effects there
(see page 221). In that case, the parallels between the two moons
would be even closer. And, for the dynamo to work, Ganymede's core
must continue to be at least partially molten today. That argues for
Ganymede having been heated up again subsequent both to its formation and to the exhaustion of radioactive sources. "Recently" (about a
billion years ago), it could easily have wandered through a high-energy
resonance that set its thermic "clock" back to zero. Did such a history
conform to the surface features now being seen by Galileo's camera?
The pictures, as Mike Belton, the scientist in charge of the camera, rejoiced, "went beyond our wildest expectations."

Ganymede in Focus: The Big Moment for Galileo's Camera

The very first pictures turned certain impressions from the Voyager
mission on their heads, and Ganymede's landscape, made up exclusively of water ice, revealed a multitude of contrasting detail. Regions
that had seemed young and smooth now showed heavy evidence of impact catering, which meant that they were old. Which meant in turn
that the grooves were definitely the more recent structures. As a whole,
the pictures themselves, despite the Sun being very high in the sky,
were extraordinarily rich in light and dark contrast. The most fascinating details in the sharp black-and-white pictures, however, came not
from the play of light, but from variations in reflectivity that represented actual differences in surface composition. Ganymede's ice, came
the quick interpretation, must have a lot of impurities in it. In those
places where the ice sublimates very effectively—where sunlight trans-
forms it directly into a gaseous state—relatively more of the mineral

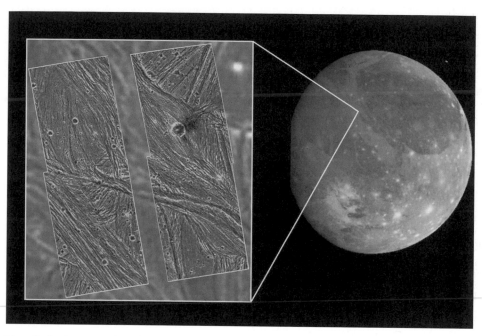

Detailed photographs of Uruk Sulcus from Galileo's first Ganymede flyby, superimposed on a Voyager 2 photograph and located in terms of the whole moon. In spite of the high angle of the Sun, the level of detail in the exceptionally high-contrast photograph was such that it overwhelmed the data-compression software, requiring a reduction in the number of scanning lines (which also caused the small holes).

admixture remains behind. Ganymede also has especially light areas, however, in very northern zones. These bright regions apparently come from frost. They are places where the water that sublimates near the equator precipitates down in the form of fresh water ice.

When exactly the moon underwent its last phase of ice volcanism has not been clarified yet. Experts are still studying the pictures, trying to answer this question. In any case, no direct signs of "recent" volcanic activity were to be seen. The most spectacular new visual evidence showed that the grooves, which are separated by several kilometers, are themselves covered by a fine pattern of ridges a few hundred meters apart which the Voyager cameras simply did not

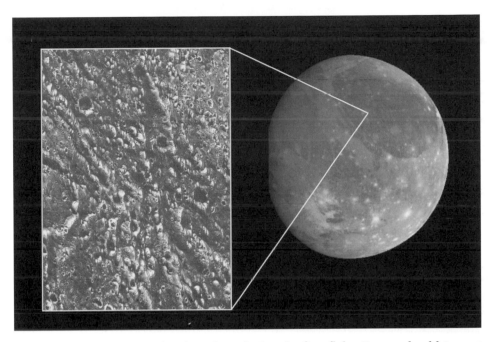

Galileo Regio on Ganymede, also taken during the first flyby. Extremely old impact craters testify to the ancient age of the area, going back several billion years.

have the resolution to see. This smaller system of grooves, with a roughly triangular cross section, can also be understood according to the Washington University model. As Ganymede expanded, causing the outer ice shell to break into the larger grooves, smaller breaks were also occurring. Ice is not infinitely elastic. Under pressure, it breaks into a fine pattern of lines to relieve the stress. The classical theory of large-scale furrowing is not affected by this detail. But scientists also admitted that the thermal and strain histories of Ganymede are certainly more complex beyond the approximations of the model. And Galileo's pictures have not made the work of the theorists any simpler, either.

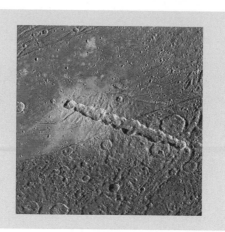

A chain of craters named Enki Catena. These 13 craters were probably formed by a comet which was pulled into pieces by Jupiter's gravity as it passed too close to the planet. Soon after this breakup, the 13 fragments crashed into Ganymede in rapid succession. The Enki Catena craters formed across the sharp boundary between areas of bright terrain and dark terrain; while the ejecta deposit surrounding the craters appears very bright on the bright terrain, it is difficult to discern on the dark terrain. This may be because the impacts excavated and mixed dark material into the ejecta and the resulting mix is not apparent against the dark background.

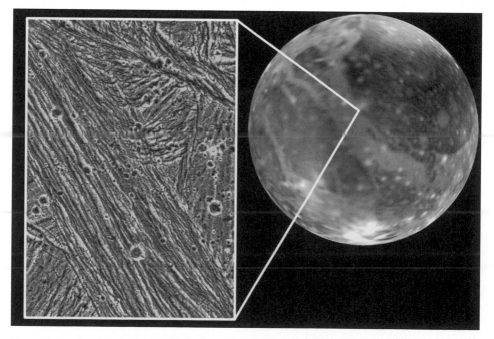

Uruk Sulcus on Ganymede. The comparison between the long-distance Voyager photographs and the Galileo closeups shows that a grooved terrain predominates in the lighter regions of the moon's surface.

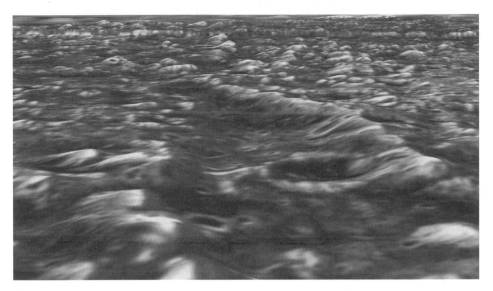

An artificially produced stereoscopic view of Galileo Regio on Ganymede. Scientists determined the topography—how the surface rises and falls—by analyzing Galileo images from two different flyovers. Deep rifts and impact craters can be made out easily.

NIMS, with its broad spectral coverage, allowed scientists to make several statements about the chemistry of Ganymede's surface. In addition to the clear signature of water ice in the spectra, a variety of minerals were also present. The water ice is spread very unevenly over the surface, playing a relatively minor role in the darker looking regions. All of these places are older, with the spectra dominated by the mix of minerals. Analyzing the NIMS spectra from Ganymede and Callisto was a painstaking job, but in 1997 it emerged that several kinds of organic molecules (cyanide, for example) are embedded in the ice. In all likelihood, the molecules are the leftover signs of comet collisions throughout the history of the Solar System. As recent laboratory tests have proved, conditions in interstellar space and on the boundaries of the Solar System where comets originate are quite conducive to the

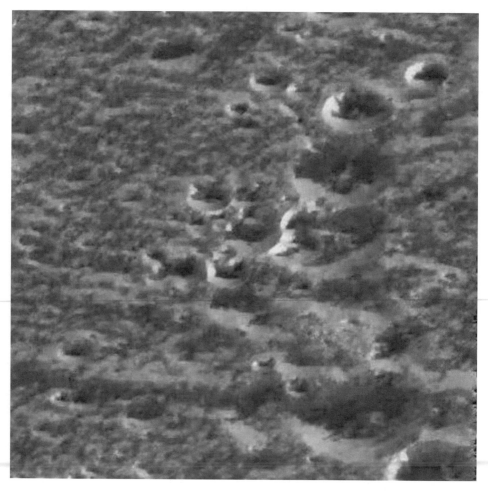

Frost on Ganymede? Scientists interpreted the traces of white in this region at 57° north latitude as precipitated frost made of water ice. Although the Sun shines from the south, the north-facing slopes and crater walls are brighter. Water molecules released near the equator by heat from the Sun seem to get caught in these cold traps.

formation of complex organic (that is, carbon-containing) chemical compounds. In the evolution of Ganymede and Callisto, these organic traces played virtually no role at all—but they would become the center of later speculation of concerning the possibility of life under the crust of the moon Europa (see page 190).

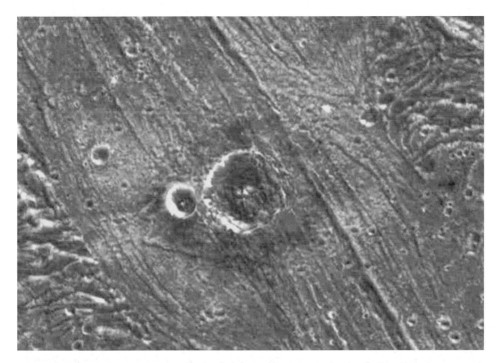

The crater Nergal on Ganymede, surrounded by unusual ejecta. Nergal's diameter is about 8 kilometers, and the other crater in the picture has a diameter of about 3 kilometers. Both are located in Byblus Sulcus, which is bright grooved terrain in the Marius Regio. The impact energy involved in the formation of Nergal obviously melted the ice in the immediate area, causing flow structures to appear and then resolidify.

The Photopolarimeter/Radiometer (PPR), measuring even longer wavelengths than NIMS, took temperature readings in various places on Ganymede's surface. Findings matched expectations in this case. The brightest regions were also the coldest, because the ground absorbs the least sunlight and reflects the most back. Even the warmest spot was only −121° Celsius. In the mornings, however, it was ten degrees warmer than predicted by simple temperature models. "A substantial revision of our understanding of Ganymede's thermal properties will probably be necessary," concluded the Galileo researchers. To them it was clear that Ganymede was not to be treated like a simple ball of ice.

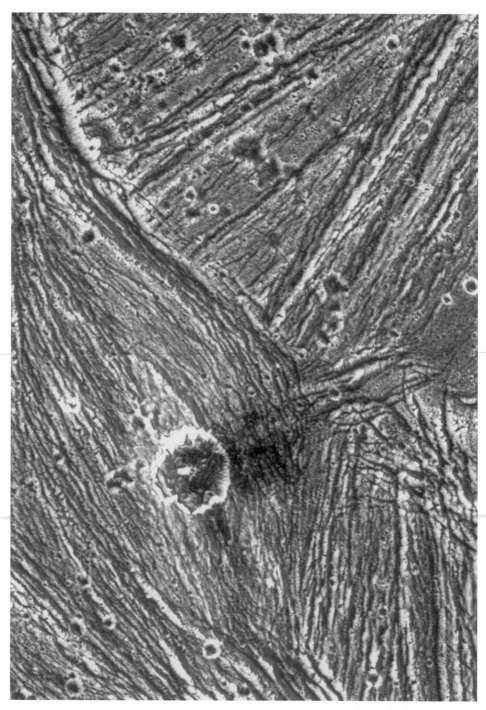

The landscapes of Uruk Sulcus on Ganymede. Heavily cratered, old terrain is present in this area along with the moon's characteristic pattern of lines. An impact crater produced dark-colored ejecta.

A number of new discoveries being made at the time, incidentally, came not from Galileo but from the Hubble Space Telescope, which sometimes took aim at objects inside the Solar System, and not only distant galaxies. Among Hubble's discoveries was the proof that ozone (an oxygen molecule with three atoms) is present in the ice on Ganymede's surface. It caused a characteristic absorption visible in the ultraviolet. Hubble also showed that Ganymede has weak polar lights in both the north and the south. Proof in this case was a characteristic *emission* visible in the ultraviolet to one of Hubble's spectrographs. Polar lights are created when magnetic fields shoot charged particles into the atmosphere (see pages 215–220 for an explanation of the polar lights on Jupiter). Ganymede, then, would have an extremely thin atmosphere made of oxygen, coming from the water ice on the surface. The existence of an atmosphere on Ganymede, which has not been proved directly, is predicated on the fact that it does have an ionosphere. From the plasma waves discussed above, it is possible to determine local electron density. Those figures proved that Ganymede had an ionosphere 1,000 kilometers above the surface, with a density as high as 100 particles per cubic centimeter. For the ionosphere to exist, there had to be a substantial atmosphere to sustain it. An atmosphere on the moon Europa is likely for the same reason, and equally lacking in direct proof (see pages 183–185).

Still more surprising was the discovery that Ganymede clearly produces abundant hydrogen. The only way this gas can occur is for water molecules to be destroyed on the moon's surface, which would leave a corresponding amount of oxygen behind. This would have to be locked in the ice—in a quantity roughly comparable to that in the atmosphere on Earth. The creation of oxygen on a solid planetary body is of great interest in regard to possible parallel processes that might have taken place very early in Earth's history.

And Ganymede had yet another surprise in store: a distinct cloud of dust, consisting of grains kicked up from the moon's surface by impacts of interplanetary meteoroids. Forty-four of these dust particles were de-tected by Galileo's dust counter during four close approaches to Ganymede in 1996 and 1997—and indications of similar dust atmo-spheres were also seen near Callisto and Europa. They all seem to be in dynamic equilibrium, meaning that as many new dust particles are pro-duced by impacts than escape into space. The escaping particles also con-tribute somewhat to the Jovian rings (which will be discussed later)—and similar processes should be at work at other bodies in the Solar System too, such as the Martian moons and even the Moon of the Earth.

Deciphering Jupiter's Colorful Clouds

The interest of planetary scientists in the Jovian system rests on two great pillars. The first is the way it can be understood as a mini solar system, with Jupiter as the "Sun" and all the moons as planets in orbit around it. The processes involved in the formation of our Solar System can be seen more clearly on Jupiter's smaller scale. At the same time, however, in atmospheric terms Jupiter is *the* most active planet in the Solar System. Whatever we might be able to understand about its weather systems could be useful in understanding the no less compli-cated atmosphere of Earth.

Nearly everything we know about Jupiter's physical nature, we have learned just in this century (as described on pages 113–120). Analysis of the chemical composition of the atmosphere, for example, began in the 1930s with the identification of methane and ammonia in the spectra of sunlight reflected off Jupiter. They are present only in small amounts, while the majority of the planet consists of hydrogen and helium. Still, not until around 1960 was there direct proof of the presence of hydro-gen. The evidence for helium had remained indirect until Galileo's at-mospheric probe finally registered it firsthand (see pages 132–134).

Sophisticated infrared detectors led to the identification in the 1970s of an assortment of molecules on Jupiter, including ethane, acetylene, water, phosphine, prussic acid, carbon monoxide, carbon dioxide, and germane. Despite being present in very small amounts, these are the ingredients in Jupiter's atmosphere which make it appear so colorful. The multiple cloud layers also contribute to the color. To understand the structure of Jupiter's cloud cover, it is necessary to analyze pictures taken in a variety of wavelengths. To include the near-infrared helps in particular. It was proved very early on, for example, that the areas that looked dark blue in visible light corresponded to the deepest layers that could still be seen. Only because there are holes in the clouds above them can they be seen at all—and the hot spot into which Galileo's atmospheric probe fell is exactly such a region. The infrared camera on the Voyager mission had peered into several hot spots and found a temperature of up to −13° Celsius. It is not possible, in other words, to look all the way down, because temperatures rise steadily with depth. Obviously, there is a defined layer in the cloud cover that maintains a temperature reading of −13° Celsius.

Hot Spots: A Glimpse into the Depths

Hot spots were also among the research objectives for Galileo's infrared NIMS instrument, beginning at the time of the Ganymede encounter. At wavelengths around 5 microns the spots represent a window into the deeper layers of the atmosphere. Some of the findings of the atmospheric probe, for one thing, could be repeated from a distance and, for another, it would be possible to make comparisons with other places on Jupiter. NIMS could "see" down to a pressure level of 5 to 8 bars, clearly registering both the absence of water clouds and the hot spots' all-around dryness. This would soon prove to be exceptional. Just as Earth has deserts and tropical regions, the "humidity" in Jupiter's clouds varies considerably from place to place. Following a global survey of

Jupiter in the infrared in mid-1997, there was no more disputing what atmospheric scientists had suspected upon receiving the first reports of surprisingly dry conditions. The probe really was "unlucky" enough to have fallen into one of the driest regions in the atmosphere. It had discovered Jupiter's Sahara Desert, so to speak.

Hot spots make up less than one percent of the surface area of the clouds as a whole. While humidity in these areas can fall as low as one percent, in other regions the reading rises to nearly 100 percent. In these places it can even rain or drizzle. Hot spots have no clouds, which allows heat from the deeper layers of the atmosphere to escape outward. They seem to be the places where wind currents come together to form downdrafts. The actual spots themselves come and go every few months. But they always appear in the same latitudinal zones, an indication that they are related to some generally stable pattern of currents. These gaseous flows caused the dryness in the exact spot the probe went down. The early alternative explanation, that all the water had gone deeper inside the atmosphere, leaving the outer regions dry, was no longer needed. The chemical composition of Jupiter was in fact turning out to be very much like the Sun's: a lot of hydrogen and helium, and only traces of oxygen, carbon, and nitrogen. These trace amounts, however, are significantly higher than they are in solar readings, which appears to suggest that they came from some later source. It may be that Jupiter has been enriched by a subsequent hail of comets.

These isolated hot spots in Jupiter's atmosphere are distinguished in the visible spectrum by a very dark appearance. Brown tones come from the next highest layers, followed by white and finally red clouds. In the infrared, the Great Red Spot is a very cold and elevated structure. The problem was that all of the cloud types needed to predict equilibrium conditions were supposed to be white. Color appears when chemical equilibrium is disturbed by charged particles, high-energy

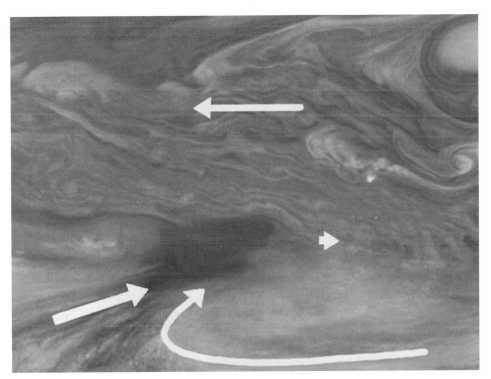

The pattern of the wind in Jupiter's equatorial region is especially evident in this picture taken in the near-infrared (756 nanometers). In the center of the picture is a typical hot spot, which looks dark in both visible and near-infrared light. The filter emphasizes the dominant cloud cover. In this image mosaic, Jupiter's northern equatorial belt (the westerly current in the upper half of the picture, with the clouds moving to the left) runs into the equatorial zone, where the atmosphere moves to the right.

photons, lightning, or rapid vertical movement through different levels of temperature. The most likely "coloring agent," meanwhile, was elemental sulfur, which comes in a variety of different colors depending on how its molecules are arranged. Parallels emerge here with the moon Io, where sulfur is also responsible for the richness of color. Other possibilities for what colors Jupiter's clouds include phosphorus and a variety of carbon compounds.

A hot spot in Jupiter's equatorial region. The picture encompasses a region 34,000 × 22,000 kilometers, extending from 1° to 19° north.

Nor was there any shortage of sights on Jupiter's nighttime side. Along with the polar lights (discussed along with magnetic fields on pages 215–220), there were also the lightning storms first observed by the Voyagers. Galileo was able to record a whole series of them, all at 44° north latitude in one of the more active regions of the atmosphere. Instruments on board the Voyagers had also registered the discharges in radio wavelengths. And yet a third phenomenon had been discovered by the Voyager cameras: meteors blazing through at about 60 kilometers per second, leaving a trail of light stretching thousands of kilometers across the sky. These were miniature replays of the fragments of comet Shoemaker-Levy 9 crashing into the planet.

Immediately apparent from the time-lapse photography of Jupiter's clouds on the daytime side is the way the wind currents flow both east and west. On Earth, the winds are arranged more simply. For each of

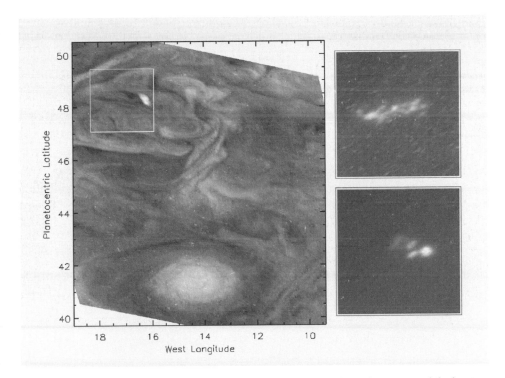

These pictures highlight a convective storm (left panel) and associated lightning (right panels) in Jupiter's atmosphere. The right images show the area highlighted (box) in the dayside view as it appeared 110 minutes later during the night. Multiple lightning strikes are visible in the night images, which were taken 3 1/2 minutes apart. The bright, cloudy area in the dayside view is similar in appearance to a region of upwelling in Earth's atmosphere. The dark, clear region to the west (left) appears similar to a region of downwelling in Earth's atmosphere. The presence of lightning confirms that this is a site of moist convection. The flashes seem to come from about 75 kilometers below the visible ammonia cloud, right from the hypothetical water cloud.

our hemispheres, there is one westward current (the trade wind) and one eastward current in the upper range of moderate latitudes (the jet stream). In both Jupiter's northern and southern hemisphere, however, there are five or six of both types of currents. In a century of continual telescopic observation, these "zonal jets" have shifted very little in latitude (or their distance from Jupiter's equator). In the four months between the two Voyager visits, wind speeds in the individual zones did not change at all. In view of the great turbulence in the winds on a smaller scale, as well as dramatic variations in the brightness and colors of the clouds, such stability is actually astonishing. The Voyagers conducted systematic observations for days at a time, however, and discovered that the winds on the large scale are moving much faster than the short-lived eddies and swirls embedded in them.

The swirls, within a week or two, are simply torn apart, transferring all of their kinetic energy ultimately to the zonal jets. The small-scale turbulence, in this way, contributes to the stability of the major currents. That the location and trajectory of the jets remain constant

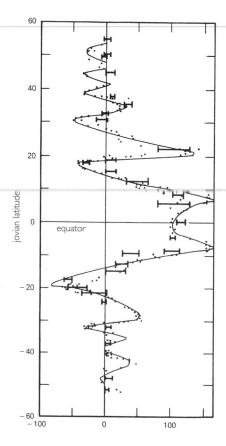

Wind velocity is strongly correlated to jovigraphical latitude. In this graph the rotational speed at various latitudes as observed from 1974 to 1979 by ground-based telescopes (horizontal error bars) is plotted against the corresponding Voyager readings. The close agreement shows that the zonal wind systems are stable over long periods of time, although the appearance of the bands and zones and the local "weather" can change drastically.

probably has to do with them being anchored deep inside Jupiter, beyond observation. Much more matter is in motion than can be observed, whether by astronomers at their telescopes or by an interplanetary spacecraft.

The stability of turbulence patterns on a very large scale on Jupiter poses an entirely different problem in dynamics. The Great Red Spot has been known for over 300 years, the three white ovals since 1939, and new whirlpools that last for several years are still being discovered all the time. Large stable ovals roll primarily along the border between opposing zonal jets, and most of them are high-pressure areas. Occasionally, two ovals will run into each other, melting together into a new whirlwind. Scientists have discussed numerous models for the Great Red Spot and other patterns like it. They could be the result of solitary waves, a type of wave phenomenon with only a single crest known from a number of branches of physics. In both computer and laboratory simulations of simplified "Jupiters," large whirlwinds develop repeatedly, melting into one another. Left over at the end is usually a miniature counterpart to the Great Red Spot.

Galileo's contribution to our understanding of these finer points, unfortunately, was limited. The loss of the high-gain antenna ruled out any systematic survey of the atmosphere as a whole. This was the most "visible" damage resulting from the antenna ordeal. What the camera was going to aim at had to be scheduled months in advance every time, a feat that could be accomplished only with constant observation of the planet from ground-based telescopes and occasional assistance from the Hubble Space Telescope. Taken together, the camera, NIMS, and the Photopolarimeter/Radiometer (PPR) could peer to different depths in the atmosphere in dozens of wavelengths both in and beyond the visible spectrum, helping scientists decipher the cloud structure. Despite the reduced transmission rate to Earth, the yield of each of Galileo's separate encounters was a data set that was simply overwhelming in its

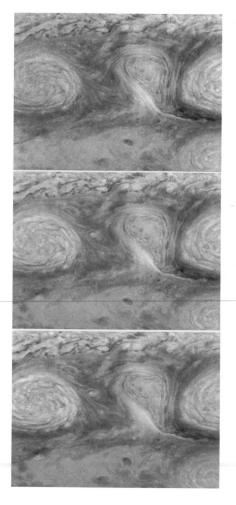

These three observations, each taken an hour apart, illustrate dynamics in Jupiter's turbulent atmosphere next to the famous White Oval Spots—two of which actually merged in 1998. Here is the situation on February 19, 1997, before the merger. Oval cloud systems like these are often associated with chaotic cyclonic systems such as the balloon-shaped vortex between the two well-formed ovals.

complexity. New methods of mathematical analysis had to be developed to deal with the multidimensional hypercubes of data. The goal was to use all the colorful new images in order to derive information about the altitude and significance of the various cloud layers.

The Spectacular Great Red Spot

The pride of Jupiter's atmosphere, naturally, is the Great Red Spot. Galileo's onboard camera and NIMS had already taken a look at it—

It all depends on wavelength. Observed in four different colors, the Great Red Spot looks different each time. These mosaics, each composed of six individual images, were produced from observations made on June 26, 1996. They show the Great Red Spot in the violet (415 nanometers; top left) and in three infrared channels (732 to 886 nanometers), all sensitive to varying levels of methane absorption, which is ideal for distinguishing cloud altitude. The highest clouds, which lie like a layer of diffuse dust over the Great Red Spot, are most visible at 886 nanometers (high methane absorption; bottom right).

from a number of perspectives and in a variety of light conditions, when the combination of Jupiter's fast rotation and Galileo's speed made it possible to do so during the Ganymede encounter. Especially obvious in particular to NIMS was how high into the atmosphere the center of the spot reaches. It remained obvious even at special wavelengths, when traces of methane in the atmosphere made what otherwise appears as the usual cloud patterns impossible to see. From NIMS data, scientists were even able to determine the altitude of the extreme upper boundary of the spot, which rises 20 kilometers into the sky, from 700 millibars at the bottom to 240 millibars. It is surrounded by a more deeply anchored ring at the bottom, and the ring in turn is surrounded by relatively clear atmosphere, making it to a certain extent like an elongated hot spot. Temperature readings showed that the center of the Great Red Spot, at −160° Celsius, is the coldest region in the atmosphere. Ammonia condenses here into rising gas currents, which turn into thick, high clouds, in a process not unlike the formation of vast water-bearing clouds extending over tropical regions on Earth. East and west of the spot, it is warmer, because the rising winds are not as strong. In the south of the planet, but very particularly in the northwest, the winds drop off, and skies are cloudless. Further analysis of the NIMS images would reveal that the GRS actually has a spiral structure of clouds, with gaps in between that allow the infrared instrument to look into the deep, relatively clear atmosphere below. Furthermore the cloud structure is higher in the center by more than 10 kilometers and tilted towards one side, like a crooked spiral staircase. What seems to be happening is that wet air from the deep atmosphere is rising rapidly in a relatively narrow region in the center of the GRS and then sprays out above the tops of the ammonia clouds like a giant garden sprinkler.

The NIMS spectra also shed new light on the nature of Jupiter's cloud layers. As expected, the main cloud layer—the one that one sees through a telescope—is made up of frozen ammonia crystals and lies at a pressure level of around half a bar. The ammonia clouds are overlaid by a thick haze at much higher levels in Jupiter's atmosphere. This appears to be a photochemical smog consisting of liquid hydrocarbon droplets. A similar layer blankets Saturn's moon Titan; although thinner than Titan's, the Jovian haze is unexpectedly substantial and varies with place and time across the planet. Jupiter's atmosphere, as outlined before, consists mainly of hydrogen, with about 15 percent helium and a number of minor constituents, the most important of which were measured and mapped by NIMS. Weather on Earth centers around the condensation and evaporation of water. On Jupiter, though, three species—ammonia, phosphine and water vapor—can condense. This makes for a remarkably complicated climate, especially since the water vapor abundance turned out to be so variable from place to place.

On the edge of the Great Red Spot, 30 kilometers above the normal cloud cover, Galileo's camera sighted several dozen bright, rapidly changing little clouds in isolation, which could soon be identified as gigantic storm cells. These towering cloud formations had been an important clue for confused scientists in 1996, when they were puzzling over the "missing" water in Jupiter's atmosphere. Only substantial

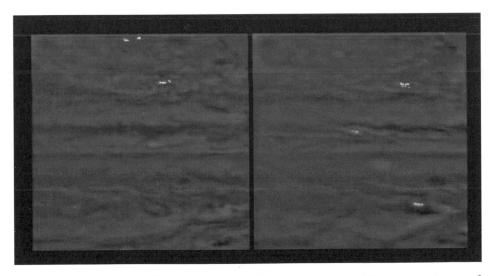

Jupiter's clouds in moonlight. These two photographs, taken 75 minutes apart and encompassing 50° latitude and longitude, offer a classic view of the clouds on Jupiter's nighttime side, illuminated only by sunlight reflected by the moon Io. More interesting are the bright spots—lightning from storms concentrated at two latitudes of particularly intense turbulence. Turbulent conditions are especially favorable for the formation of storm cells, just as they are on Earth. Jovian storm clouds produce lightning at approximately the same rate we observe it on Earth, but the lightning bolts on Jupiter are several orders of magnitude brighter than they are here.

quantities of water could explain these explosive convection phenomena, because other gases on Jupiter lack the necessary energy. A complete picture of the vertical atmospheric currents was now taking shape. As a result of falling into a hot spot, the probe had ended up in a powerful downdraft of the kind that also occurs over deserts on Earth. These currents are dry because, as updrafts in neighboring regions, they have already had the humidity stripped out of them. The probe had simply missed the water—and Jupiter could return to being thought of as a planet full of stormy weather, lightning bolts, and gales. If the behavior of these towering clouds has now finally been understood correctly, according to meteorologists on the Galileo team, it could help improve our weather models on Earth. In this way, humanity as a whole can benefit from observations in distant space.

Progress Report: What Powers Jupiter's Clouds

Slow but steady progress has been made in understanding how Jupiter's weather works. As Galileo's images—few as they are—reveal, there is a clear association of lightning storms on the giant planet with the eddies that supply energy to the large-scale weather patterns. Such a conclusion is possible because Galileo can provide daytime photos of the cloud structure when lightning is not visible, and nighttime photos of the same area a couple of hours later which clearly show the lightning. From the nighttime images even the depth and intensity of the lightning bolts can be determined. Especially fortunate are such Jovian nights when there is a bit of moonshine from one of the large moons such as Io. When the upper clouds are illuminated just a bit, both the lightning flashes and the atmospheric eddies they are associated with show up in the images. For a long time Jupiter researchers have known that Jupiter had lightning, and since the Voyager flybys it was also clear that the zonal jets and long-lived storms are kept alive by soaking up the energy of smaller eddies. But the new insight from Galileo's images is that the eddies themselves are fed by thunderstorms below them.

"The lightning indicates that there is water down there, because nothing else can condense at a depth of 80 to 100 kilometers," explains planetary atmospheres specialist Andy Ingersoll. "So we can use lightning as a beacon that points to the place where there are rapidly falling raindrops and rapidly rising air columns—a source of energy for the eddies. The eddies, in turn, get pulled apart by shear flow and give up their energy to these large-scale features. So ultimately, the Great Red Spot gets its energy and stays alive by eating these eddies." Adding credence to the interpretation is the fact that the anticyclonic rotation of the eddies (clockwise in the northern and anticlockwise in the southern hemisphere) is consistent with the outflow from a convective thunderstorm. Their poleward drift is consistent with anticyclones being sucked into Jupiter's powerful westward jets. Interestingly the physical attributes of Jupiter's vast thunderstorms are the same as those on Earth—only that Earth's storms develop because of the Sun's heat, while Jupiter's storms develop from the planet's own internal heat source. But the basic process is the same: "moist convection" in which water condensation is one of the main sources of warming the air which then rises.

An Ocean Under Europa's Icy Crust?

Our understanding of no single aspect of the Jovian system has been enhanced more by Galileo than the moon Europa, located in space between Io and Ganymede. A low density of 1.9 grams per cubic centimeter made it easy to determine that Ganymede and Callisto were made of ice. Density values for Io and Europa, however, came in at 3.5 and 3.0 grams per cubic centimeter, respectively—a clear indication of the substantial presence of rocky material. Prior to 1979, most planetary specialists had therefore assumed that Io and Europa would look much like our Moon or Mercury, little more than stone deserts. Already before the Voyager trip, telescopic observations had shown that Europa, with a diameter of 3,130 kilometers, had a crust made completely of ice. Again based on density calculations, the ice appeared to be as much as 100 kilometers thick. But no one guessed even remotely what landscapes had been created by this ice. Voyager 1 had come only to within 734,000 kilometers of Europa. From this distance, the moon looked like a very bright, smooth ball of ice—but it was covered by a thick network of dark lines, some of them 3,000 kilometers in length.

Voyager 2 pictures, which were four times as sharp, only sparked more confusion. Aside from the lines and dark spots in the ice, there

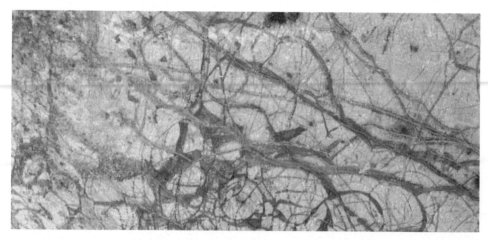

The plates into which Europa's ice crust is broken are as much as 30 kilometers across, with dark material having risen up between them. The dark color is most likely due to mineral contamination. The overall image of the dark zones and the lines in between the ice plates immediately reminded many planetary scientists of the historical renderings of the "Martian canals" 100 years ago.

were very few impact craters to be seen. Either the surface was very young, or the craters disappeared in the ice by themselves. Image analysts took the dark lines at first for cracks in the ice, but that conclusion had to be revised. There are no depressions in the ice, for Europa has no topography. Describing such a world sent scientists searching for analogies; for example, "a billiard ball marked with a felt tip pen." But there were also light-colored lines, which in oblique sunlight proved to be the crests of mountain ranges several hundred meters high. And they were not straight, but formed in arcs. "The impression

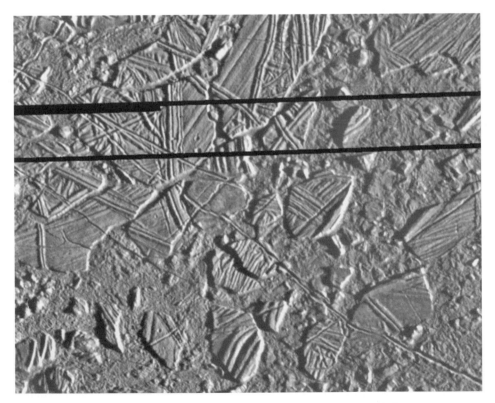

Europa's plates: the ice crust has broken up into plates up to 13 kilometers across, which have then shifted in relation to each other—very much like what we know from the polar seas on Earth. But the plates did not necessarily have to be floating on liquid water to shift like this—just barely frozen ice would have been sufficient.

is so bizarre that one tends not to believe the reality of what is seen. Nothing remotely like it has ever been seen on any other planet," NASA scientists said in a 1979 book on the Voyager discoveries. All kinds of parallels to the pack ice in the Earth's polar regions were proposed, but fresh insight into Europa's geology would have to await Galileo.

During the first Ganymede encounter in June 1996, the camera had also ventured a distant look at Europa. Even these early pictures represented something of an improvement over the best Voyager images. They were ready for release in early August—and the timing could not have been better. News of the discovery of possible signs of life in a meteorite from Mars had just exploded in the media. The next round of speculation about life in the Solar System would not be far behind, because even these distant Galileo photographs of Europa's icy surface suggested that the moon is at least partially melted! Nor did it take long for thoughts to leap from a hypothetical ocean underneath the ice to the possibility of a life form being found in the water. Given the fascinating pictures, experts were not always able to resist speculating. The long dark lines had white stripes down the middle—earning them the name "triple bands" for the dark–light–dark alternation. They were revealed to be gaps between individual plates of ice, which, however, have been filled up with some darker material. It seems that dirty water has come up from springs and solidified.

Scientists interpret the shape of the plates themselves as an indication that the medium underneath them is at least viscous, either "warm ice" at about the melting point or actual flowing water, which in our Solar System is an absolute rarity. Every one of Galileo's flybys brought new discoveries, as well as ever sharper and more spectacular images. Soon, among the experts putting in an appearance at NASA's always-popular press conferences were polar scientists and even ocean specialists, a unique event in the history of planetary research. Europa's most remarkable landscapes were these severely fractured and relatively young ones, where there were genuine ice floes. Now, with the improved resolution of Galileo's pictures, they bore an even greater resemblance to our own polar seas in springtime. Because the floes had broken off from larger areas that were covered with grooves, it was possible to reconstruct their past movements in detail. Computer models already exist in which the motion of the fragments can be traced back into their original positions. A number of possibilities are left open, however, as to what it is underneath that they are sliding on. The models work with water, but warm ice is also possible. They do not yet represent any clear proof of an ocean.

Another Planetary Ocean in the Solar System?

The ice floes rise 100 to 200 meters above the surface, and since—as is true of icebergs here on Earth—90 percent of the volume lies beneath the surface, the floes may be a kilometer or two thick. Europa's crust, then, would probably not have been thicker than that in those spots where the water under the surface had sometime become liquid or viscous, with the resulting currents then causing a certain amount of fracturing to take place. Should it be the case that Europa is still completely liquid beneath the surface ice, the ocean would have depths from 100 to 200 kilometers and contain more liquid water than all the Earth's oceans combined. It would be the first discovery of a "planetary ocean" in 500 years, since Magellan circumnavigated the globe. The heat once responsible for melting Europa's ice was probably generated by tidal forces that kneaded the moon during one specific phase in the development of its orbit around Jupiter. It must be the same process that probably affected Ganymede sometime in the past, and which continues today on Io. Tidal forces caused the most deformation on the ice crust of Europa's ocean, which in most places is about 10 to 30 kilometers

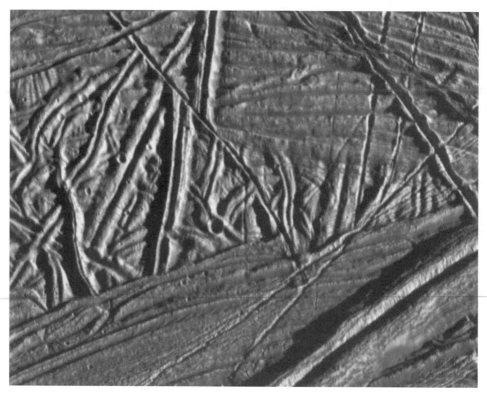

The surface of Europa is covered with multiple ridges and fault lines, as shown in this 15 × 12 kilometer detail. The prominent ridge at the bottom right measures about 2.5 kilometers in height and is one of the youngest structures pictured here—it cuts across many of the other formations.

thick, and had a less severe effect on the moon's rocky mantle underneath the ocean. The conspicuous network of lines covering Europa's surface may have been created in one such tidal episode. Forces must have deformed the moon's surface to such an extent that they broke it up into innumerable "continental plates."

At that point, "dirty" water must have risen up into the cracks and become frozen, creating this strange pattern of more and less prominent dark spots and lines, ultimately resulting in the "triple bands."

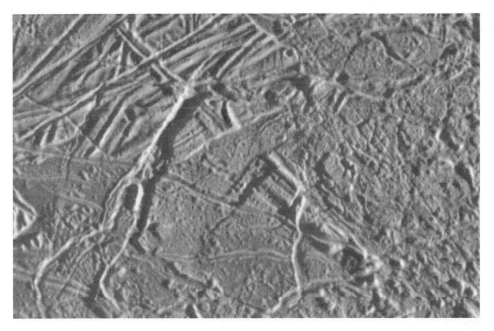

One of the highest resolution pictures of Europa, encompassing just 10 × 16 kilometers. It shows the complex pattern into which the moon's ice crust has broken. Also visible are several small craters, 100 to 400 meters in diameter.

Detailed models exist of how the water would have frozen in between the plates and then been pushed up by internal pressures to form the slightly raised bands—always based on the assumption that there is liquid water underneath the ice.

An impact crater called Pwyll also fits into this scenario. Pwyll is perhaps only a million years old, and at 25 kilometers in diameter it is the only large crater on Europa. A close look reveals the ejecta thrown up by the crash to be surprisingly dark. Did the meteorite break completely through the surface? Were the dark traces left by "dirty water" that splashed up from the depths and immediately froze? Other features of Europa's surface include ice volcanos, which periodically spray parts of the surface with icy lava, leaving behind remarkably

Europa's surface on the scale of the San Francisco Bay. The two pictures are the same size, 13 × 18 kilometers. The Galileo pictures show both furrowed and smooth regions where the unevenness has been partially ironed out by "lava." The photograph of Earth was taken by a Landsat satellite.

This complex area on the side of Europa facing away from Jupiter shows several types of features which are formed by disruptions in Europa's icy crust. The prominent wide, dark bands are up to 20 kilometers wide and over 50 kilometers long. They probably formed when Europa's icy crust fractured, separated and filled in with darker, "dirtier," ice or slush from below. A relatively rare type of feature on Europa is the 15 kilometers impact crater in the lower left corner. A region of chaotic terrain south of this crater contains crustal plates which have broken apart and rafted into new positions; some of these "ice rafts" are nearly 1 kilometer across. The younger features in this scene are the long, narrow cracks in the ice which cut across all other features.

smooth surfaces. Unfortunately, the dating of Europa's various surface structures remains in dispute. The dramatic transformations may have taken place hundreds of millions of years ago, or they could be just a few million years old. And, finally, there is the possibility that Europa remains active today.

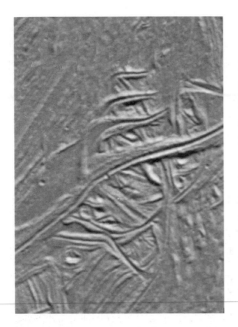

These cracks and ridges in the south polar region of Europa have been rotated into sigmoidal or pS-shapes by the motion of Astypalaea Linea, a strike-slip fault in the moon's icy surface. These cracks and ridges are located within Cyclades Macula, a region of the fault which has been pulled apart. The openings created allowed warmer, softer ice from below Europa's brittle ice shell surface—or water from the possible subsurface ocean—to reach the surface. This upwelling of material formed large areas of new ice within the boundaries of the original fault.

The global system of raised ridges provides geologists with the clearest indication, even more compelling than the ice floes, that there is liquid water underneath the surface. The argument they make is a subtle one. The fact that the pattern of breaks in the surface is global implies that the ice crust was completely uncoupled from the moon's rocky core when the fracturing occurred, and may continue to be so today. The rotation of the core is such that it has one side turned permanently toward Jupiter. Small irregularities in the distribution of Europa's mass must be responsible for this "bound rotation," a phenomenon we find everywhere in the Solar System, and which in fact describes the rotation of our own Moon. Europa's ice crust, on the other hand, rotates a bit faster than the bound core, suffering severe stress as a result, because Europa as a whole is not an exact sphere. Scientists reason that the fractures in the crust provide indirect evidence of the existence of an ocean in between the crust and the core, because only a liquid could accomplish the required uncoupling.

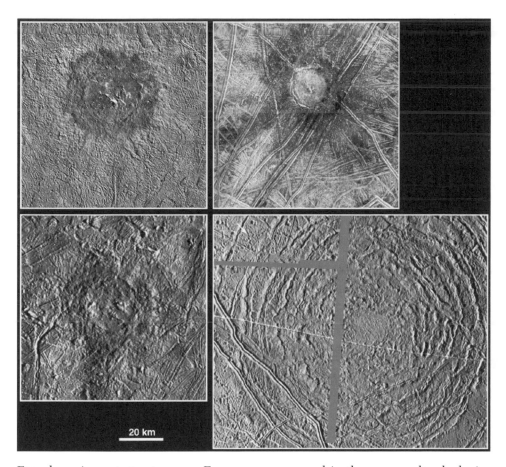

Four large impact structures on Europa are compared in the same scale: clockwise from top left they are Pwyll, Cilix, Tyre, and Mannannan. Impact structures with diameters of more than 20 kilometers are rare on Europa, and Tyre is the most unusual case. While the effective crater—somewhat larger than the prominent bull's eye feature—measures about 40 kilometers, the entire structure is much larger. The concentric rings display relatively little relief. Some of the smaller craters near Tyre were formed by material ejected by and redeposited from the impact which formed Tyre—the impactor may have penetrated through the icy crust into a less brittle layer. While Pwyll, Cilix and Mannannan also display shallow crater depths for their size, they more closely resemble similar sized craters on Ganymede and Callisto. Perhaps the impactor did not punch through the upper crust during these events.

While there remains little doubt of an ocean having existed at some point in Europa's *past,* in other words, a clear indication that it continues to exist is lacking. Perhaps, in a possible interpretation of the Galileo images, part of the crust is resting solidly on a rocky pedestal, while other regions (especially those where the "icebergs" have been observed) really do have water underneath. In this way, the existence of a liquid ocean today can be reconciled with the observation that some areas of Europa's surface seem to have been inactive for tens of millions of years, as evidenced by a collection of impact craters.

Galileo found signs of a magnetic field around Europa, as it did for Ganymede, but only about one quarter as strong. And the field's axis is tilted an astounding 45° to Europa's own rotation axis. As of yet, there is no clear analysis. Readings of Europa's gravitational field revealed, as they did for Ganymede and Io, that it is a differentiated body, with a metallic core surrounded by a structure of layers, much like Earth. Data from four close flybys of Europa (during the orbits 4, 6, 11, and 12) have allowed researchers to paint an ever clearer picture of Europa's interior. There is, in all likelihood, a metallic core which could have up to 50 percent of Europa's radius, depending on its exact chemical composition. The next layer is a rocky mantle, and on top comes a shell of water and/or water ice, with a thickness of 80 to 170 kilometers. Gravimetry cannot tell, unfortunately, whether there is liquid water beneath the ice, because the densities of water and ice are too similar. And the three-component model of Europa outlined above is not totally unique either. There remains a marginal possibility that underneath the ice there is a uniform mixture of dense silicates and metal. Confirmation of the magnetic field would probably argue in favor of the metallic core model. Europa would be a second Io to a certain extent, but with an outer covering of ice. Unfortunately, the big question, whether part of the ice is melted, cannot be answered by further analysis of the gravitational field.

An Atmosphere of Its Own?

Europa has yet another obvious feature that kept scientists busy. It has a thin ionosphere, which means that it has an atmosphere! Prior to Galileo, there had been proof of a thin atmosphere only on Io, but predictions had been made for Europa. It was suspected that the bombardment of the ice crust by particles would cause water molecules to be released, which theoretically would lead to an atmosphere consisting of water vapor, monatomic and diatomic hydrogen, hydroxyl (OH), and monatomic and diatomic oxygen. But Galileo found nothing, and only in 1995 did ultraviolet spectra taken from Hubble allow astronomers to make a breakthrough, after 23 years of futile searching. The spectrograph, after a two-hour exposure on distinct emission lines, had detected atomic oxygen, although the atmosphere probably consists mainly of molecular oxygen. The discovery of an atmosphere was responsible for a certain media stir in early 1995, but it was so thin that the excitement quickly subsided. Pressure on the surface would amount to just 10^{-11} bar. Europa's atmosphere extends perhaps as far as 200 kilometers all around it, but condensed to the density of Earth's atmosphere, it could fill only a dozen sports arenas.

The origin of Europa's atmosphere is meanwhile relatively well understood, with a total of three different forces involved in the release of water atoms from the surface ice. The ice itself sublimates some water in the sunlight—very much like what happens to the core of a comet. Bombardment by dust particles knocks loose a few more fragments. By far the most important process, however, is called "sputtering." Charged energetic particles from Jupiter's magnetosphere hit the surface, knocking water molecules free. Hydrogen, being very light, escapes, while the oxygen lingers a bit longer near the moon. Electrons from Jupiter's magnetosphere collide with it there, splitting O_2 into two O's, and at the same time causing the atoms to glow. This process indirectly produces the ultraviolet radiation that was detected by the Hubble telescope.

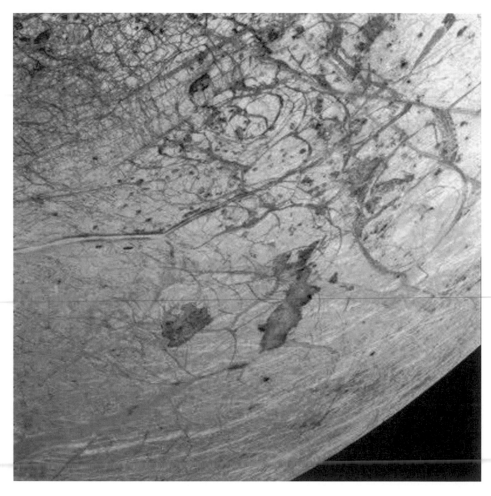

Thera and Thrace Macula in Europa's southern hemisphere are visible as two dark spots in this photograph. To the north is the Agenor Line, which is almost 1,000 kilometers long.

Oxygen is also constantly escaping Europa's gravitational field, whether because of the thermally induced motion of the molecules themselves or by other mechanisms. Particles from the magnetosphere make yet another contribution to the erosion of Europa's atmosphere. Bombardment by the particles, like ultraviolet radiation from the Sun, robs the oxygen of electrons. The electrons, in turn, make up the ionosphere, which Galileo had detected in 1996 and 1997 with radio science. A series of three scans in space around Europa linked all the evidence together. Curiously, however, Europa's atmosphere, according to figures derived from this evidence, would have a temperature of 70° to 300° Celsius—much higher than that of the ice below it. Jupiter's magnetosphere must also be working as a heater for Europa's atmosphere. A by-product of the radiation bombardment of Europa's surface is the formation of hydrogen peroxide—the chemical well known to turn a brunette into an instant blonde. Its signature was detected by the NIMS instrument in 1999.

And this is not all. In 1995, observers using ground-based telescopes detected traces of the chemical element sodium in Europa's atmosphere, since astronomers are able to register the emission line of sodium optically. It really has no business being in Europa's water ice, so it may be a souvenir from one of Io's volcanos, which are known to emit sodium. Particles from the magnetosphere around Io collide with this sodium and ionize it, effectively trapping it in Jupiter's magnetic field. At that point a certain amount of the sodium is shot at a very high speed directly into Europa's ice. There it becomes subject to the same processes that cause the decay of water molecules, and finally it ends up back in the atmosphere. This is probably the only example in the entire Solar System of a moon's atmosphere being fed, as it were, by one of its neighbors.

Europa's Crust: Young or Old?

Only rarely were Galileo's findings about Europa so definitive. The question of what goes on under Europa's crust is not the only matter that continues to stir astonishing levels of controversy. Also in dispute is the age of the crust itself, with estimates in 1997 ranging from a few million to several billion years—which means they differed by a factor of ten! Determining the age of a solid planetary body is usually relatively

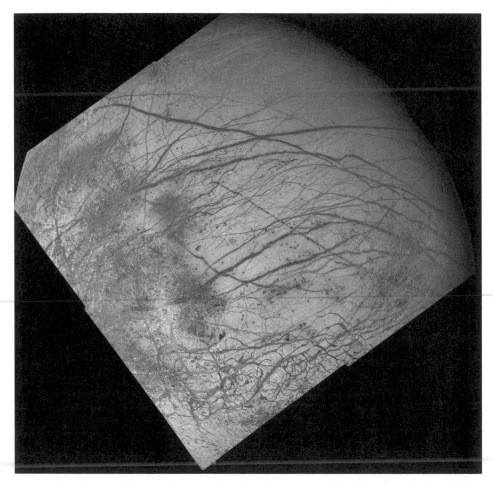

This early mosaic of four photographs taken from 156,000 kilometers in June 1996 shows a large expanse of Europa's fascinating surface. Most striking are the dark lines, some of which are over 1,600 kilometers long. At the very bottom is a region where the ice crust has been broken into many small plates that are a maximum of 30 kilometers across.

straightforward. Astronomers record the number of impact craters they can identify over a certain surface area, deriving the age of the body directly from the frequency of impacts.

But there are at least a couple of problems with this technique. The only way astronomers can "calibrate" the scale is to use what they

A Europa photograph taken on November 6, 1997, during Galileo's final visit of the primary mission. It shows the mottled terrain characterizing the surface where the most recent known geological activity on Europe took place. Scientists speak of "chaos terrain" in reference to this area.

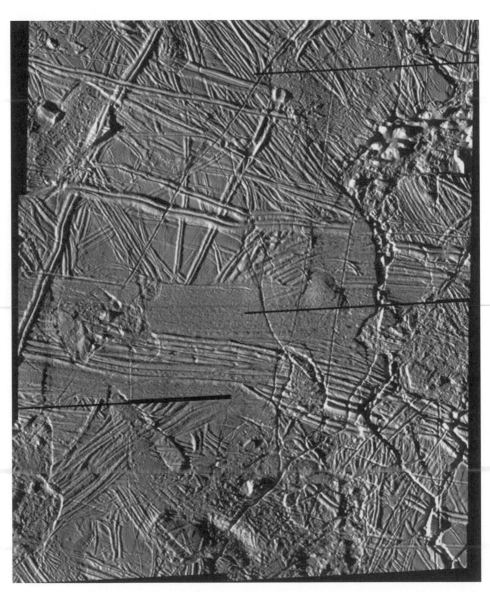

One of the highest resolution Europa pictures of the primary mission, from the final flyover on November 6, 1997. Traces have been left here of several episodes of geological activity. Also visible are several isolated mountains rising to a height of about 500 meters.

know about the rate of crater formation on our Moon—the precise age of which is known from the direct analysis of rock samples brought to Earth by the Apollo astronauts. Still, transferring the crater formation rate in this part of the Solar System to the Jovian system required a lot of adjustments. Not even the current rate of crater formation is very well known for Jupiter's moons, and it could have been different in the past. It is precisely this assumption, that earlier in the Solar System the rate of collisions must have been much lower, that leads a minority of planetary specialists to estimate the age of Europa's surface at over a billion years. In that case, the moon would have been dead for eons by now, and there would be no chance of an ocean existing there today.

Most Galileo researchers believe they have the mathematics well enough in hand to guess the age of the youngest regions of Europa's crust, the "ice floe" fields, at 1 million to a maximum of 100 million years. Why would the process that set the ice floes in motion, causing them to turn at angles to each other, go on until just a few million years ago and then suddenly stop? High-resolution photography of broad stretches of Europa's crust, which Galileo delivered during the extended mission from 1997 to 1999, could clarify the age question. Every additional crater count increases certainty. Moreover, crater chains, the impact sites left by fragmented comets like Shoemaker-Levy 9, have already been observed on Callisto and Ganymede, and finding them on Europa would finally make it possible to determine how many small comets there are in Jupiter's vicinity. Meanwhile new numerical simulations of the processes that "deliver" both asteroidal and cometary bodies to the Jovian system continue to clarify the picture. By 1998 it seemed clear that 90 percent of Europa's craters were produced by comets, hardly one was attributable to an asteroid. This has led to the most precise age estimate for Europa's surface so far: it is between 2 and 50 million years old—in geological timescales this is now. And the youngest regions of the surface may have an age of just 500,000 years.

Europa gives scientists also a good opportunity to study how the ejecta (material ejected when large craters are formed) goes on to produce many smaller secondary craters. At least two-thirds of the small craters on Europa are secondary craters left from the Pwyll collision.

The Question of Questions: An Ocean with Life?

The possibility that there is an ocean under Europa's crust took hold in the public awareness in 1996 and 1997 like no other Galileo discovery. NASA planetary experts were soon thrashing out on the public stage whether the most interesting planetary body was Io, with its volcanos, or Europa, with its possible ocean. Science fiction fans naturally knew the answer from the start. In his book *2010*, Arthur C. Clarke had picked up on earlier speculation, right after Voyager 2, about an ocean on Europa, and had planted a mysterious life form there. For humanity to study it was taboo.

Things turned out differently in reality. Galileo's cameras cannot peer beneath Europa's ice, and speculating about what goes on under there is irresistible. The reason has to do with another discovery, this one made a good two decades earlier in a completely different part of the Solar System—the ocean floor on Earth. Small maneuverable research submarines had just become available, allowing scientists to discover hot volcanic springs at the bottom of the ocean. These springs are surrounded by animal colonies, harboring species that have never been seen before. The remarkable thing was that these life forms were flourishing *in the complete absence of sunlight*. Prior to this discovery, biologists had assumed that ultimately all life on Earth must tap directly or indirectly the energy of the Sun. All known deep sea life forms were ultimately nourished, it seemed, from sinking biomass that originated closer to the surface, where sunlight still penetrated. The research conducted with small submarines along the mid-Atlantic ridge changed that view completely.

That is where researchers came across the famous black smokers, hot undersea chimneys that exude a blackish liquid that is very rich in minerals. Fish, absolutely colorless crabs, and tube worms 3 meters long flourished around the smokers, which were obviously what made life possible in that environment. Then scientists made an even more fundamental discovery. Microorganisms lived in the hot water, heat-tolerant bacteria for which 55° Celsius is the optimum temperature. And "tolerant" can be something of an understatement. A species from the Azores region starts growing only at 95° Celsius and easily endures temperatures as high as 113° Celsius.

Since the surrounding water is only 2° to 4° Celsius, these bacteria cannot have come from there. They came from inside Earth. This realization transformed the biologists' whole viewpoint. Suddenly it was necessary to consider whether a substantial portion of the Earth's bio-mass might not be located underground. Not much later, microbes were discovered in rocks at depths of 2.8 kilometers, living in tiny cracks in the granular matter, which the microbes consume for nourishment.

Why would it not be possible for this phenomenon discovered on our own ocean floor to be taking place somewhere else in the Solar System? Seen in this way, the probability of life under Europa's ice crust was even greater than that of life on Mars, which had always been the prime candidate for speculation concerning extraterrestrial life forms.

Tidal forces exerted by Jupiter and its moons—a source of heat that may have melted Europa's ice—could just as easily be responsible for the existence of undersea hot springs. At least that is the result of thermal models that have been made of Europa's possible development. Comets crashing into the surface would have supplied plenty of the basic building blocks of the chemical compounds that are essential for life. On Ganymede and Callisto, NIMS had already found traces of organic compounds (as discussed on pages 153–154), a discovery that encouraged optimism about the possibility of life on Europa. The principal ingredients of life—water, chemistry, and heat—could be present there. Whether that suffices for a primitive life form to emerge, however, is an entirely different question, and it cannot be answered in the absence of clear examples from somewhere other than Earth. The enthusiasm for Europan life was dampened in 1999 when biologists looked at the chemistry in greater detail. Apparently the Black Smokers in Earth's oceans are only so hospitable for life because they are constantly supplied with oxidants from the upper oceans—there is no such supply on Europa and has never been. But perhaps life has found alternate means. And speaking in favor of the possibility of life elsewhere is the speed with which it did emerge on Earth—within a few hundred million years of the planet first becoming habitable about 4 billion years ago. In a permanently liquid ocean on Europa, life would have had all the time in the world to develop . . . .

NASA press conferences sometimes made it seem that the necessity of life developing on Europa had been established as a fact—assuming, of course, that the evidence had been correctly analyzed. In August 1996, at the first presentation of the now vigorously disputed "Mars fossils" to the public, NASA head Dan Goldin had succumbed to dramatic overstatement, and now he did it again. The images were "exciting and compelling, but not conclusive. The potential for liquid water on Europa is an intriguing possibility, and another step in our quest to explore the Solar System, the stars and the answer to the great mystery of whether life exists anywhere else in the cosmos." The U.S. news media, in particular, were only too happy to play along. "Perhaps the most significant discovery in the annals of space exploration," was how Ted Koppel put it on *Nightline*. He expressed the hope that, in the foreseeable future, the "perhaps" could finally be left out of statements about the existence of life elsewhere in the universe.

That day might be a long way off yet, but at least the study of Europa is nowhere near its end. The treasure trove of data from the Galileo Europa Mission from 1997 to 1999—with the majority of the flybys successful—awaits detailed analysis. Despite intense searches for changes on Europa since the Voyager 2 visit no indications for current geological activity has been found by early 2001. On the one hand we have the extremely young age of parts of Europa's surface told to us by the crater counts, on the other hand there are no ongoing changes. Could it be that the moon froze just "recently"? This is one of many questions studied now by theorists. Perhaps Europa's interior melts only periodically when the constellation between the Galilean satellites of Jupiter leads to particularly pronounced tidal heating. Then we would have bad luck with our search for current life on the moon. Other theoretical work centers on the mechanisms that have formed the various surface features—and on the basic question of whether life is possible at all under Europan conditions. The suspense over whether life exists in any form on one of Jupiter's moons will be with us for a long time yet.

Progress Report:
An Ocean on Europa? The Clearest Evidence

In the spring of 1998, the discussion for and against the existence of an ocean on Europa was still going full tilt. The alternatives continued to be posed in terms of completely fluid water on the one hand, and viscous ice on the other. The only undisputed statements were the following: the outer 150 kilometers of Europa consisted of H_2O, in one form or another, and this water was at least not in a completely frozen, rock-hard condition. Also beyond dispute was the amount of water involved, which could be derived with relative certainty from data on Europa's gravitational field and models of its interior structure. But now scientists were able to prove that the rocky interior of Europa was not firmly attached to the outer crust.

Although the interior of the moon always maintains the same orientation to Jupiter, locked with it in a "bound" rotation, the ice crust evidently does not. Very slowly, the part of the moon's surface that is turned toward the planet is changing, so that a complete revolution will take tens of thousands to tens of millions of years. The clues that the core and the crust are decoupled were subtle, but they became increasingly persuasive over time. It is possible to determine from an analysis of Europa's gravitational field that the core is bound exactly to the rotation of Jupiter. Evidence of the asynchronous rotation of the crust is found in the direction, number, and relative ages of the long fissure lines covering Europa's surface. Also clear is the presence of some kind of "lubricant" between the core and the crust—but it could just as easily be fluid water as viscous ice.

From the remarkable closeups of Europa's crust, scientists were able to distinguish four fundamental components of the landscape: "background" plains, lines, areas that had undergone local disruptions, and areas bearing signs of a major disturbance (as in the photograph on page 195). The plains are the oldest formations, consisting of a densely woven network of intersecting ridges. Then came the long "lines," some of them extending for thousands of kilometers. They are dark-colored ridges rising parallel to each other to a height of 100 to 200 meters. In terms of the existence of a subsurface ocean, however, the most interesting landscape occurs where the plains have been deformed, either because of local forces attending the formation of 10- to 20-kilometer-wide bulges, or on a large scale by the "ice floes" drifting apart.

The local variations seem to exist everywhere on Europa, visible as dark spots even on low-resolution photographs taken from very far away. Most dramatic looking are the areas where the crust has been severely disturbed—such as the diamond-shaped region in the photograph on page 195. The crust has been broken into pieces ranging up to 20 kilometers across and towering 100 to 200 meters over their surroundings.

Geologists regard this mosaic, composed of six individual images from the sixth orbit and encompassing an area 300 × 300 kilometers in size, as the most important of all. The landscape to the south of (below) the large cross, with two prominent line systems, or "triple bands," is conclusive. The crust here has been broken into blocks that have moved laterally from their original positions, as can be seen in the patterns on them left over from earlier times. Outside the 75 × 100 kilometer zone in which geological disturbance is evident in the mosaic are a variety of flat bulges, from several kilometers to nearly 20 kilometers in diameter, photographed with a resolution of 180 meters per pixel. They are a sign of subsurface thermal activity.

From the old patterns on the otherwise smooth tops of these giant ice blocks it is evident that some of them have traveled a distance of several kilometers and have also rotated relative to each other. And since Europa as a whole is extraordinarily flat, with no slopes of any kind, the shifting cannot have been caused by gravity. Lateral movements must be a result of the movement of the material in which the ice plates are embedded.

As the plates drifted, the underlying material must have flowed upward into the cracks opening up behind the plates—and portions of the ice crust *in front* of the plates must have been submerged or destroyed. Yet, that is the end of what is known for certain about Europa. It cannot be determined from the photographs whether the breaking up of the ice was a relatively rapid process, with the plates floating on liquid water, or a slow one, with the rigid plates sliding on viscous ice that gradually rose up into the gaps between them. Would the answer be found in the most detailed of all of Galileo's photographs, taken in December 1997? Scientists presented the photographs to the public on March 2, 1998, citing three indications in particular that under Europa's crust even today there is either liquid water or, at the least, very fluid ice.

First comes Pwyll, an impact crater only 10 to 100 million years old which is extraordinarily flat. The basin caused by the collision must have filled up again immediately—with a liquid from under the surface. Evidence that it was at least partially melted ice—slush—comes from what image analysts term "chaos terrain," detailed photographs of which have now become available. Rough, twisted-looking structures in between individual ice floes in numerous areas are suggestive of frozen slush. The dark wedge-shaped zones between Europa's larger ice plates can also be studied now in unprecedented detail. For planetary geologists, their parallel ridges and grooves are strongly reminiscent of the new formation of Earth's crust taking place on the mid-Atlantic ridge. In what amounts to an initial summary of findings from the high-resolution photographs, this is all evidence pointing to the existence of liquid or partially liquid water just beneath the surface of the moon in the very recent past.

A major discovery about the workings of Europa's surface was made in 1999. Geophysicists playing around with models of how tides affect the moon's icy crust hit upon a convincing explanation for one of the most mysterious features seen in the ice. The phenomenon had already been discovered in Voyager 2 images: so-called *cycloids* or *flexi*, chains of arc-shaped ridges extending over hundreds of kilometers which are found all over the moon. The new model attributes the feature to the diurnal variation in tidal stress in Europa's outer ice shell. Because Europa's orbit is elliptical, there is a tidal bulge wandering around its body, a deformation severe enough to crack the surface under certain conditions.

When the tensile strength of the ice is reached, a crack could occur— and it would propagate "across an ever-changing stress field, following a curving path to a place and time where the tensile stress was insufficient to continue the propagation," explain Gregory Hoppa and colleagues from the Lunar and Planetary Laboratory. "A few hours later, when the stress at the end of the crack again exceeded the strength, propagation would continue in a new direction. Thus, one accurate segment of the cycloidal chain would be produced during each day on Europa." It would be fun to watch such a cycloid forming, almost racing over the surface.

And there is more to the story. For this clever model to work, "there must be a thick fluid layer below the ice to allow sufficient tidal amplitude"—the surface must be able to change by about 30 meters over 43 hours. Were Europa's water layer solid down to the silicate core without a substantial liquid layer in between, tens of kilometers thick, the diurnal stress would be inadequate to fracture the surface because the amplitude of the tidal variation would be so small. Hoppa and his team conclude: "The creation of cycloidal cracks requires a global ocean"—one of the strongest, if indirect, arguments for an ocean based on surface features proposed so far.

Other indications that an ocean continues to exist today on Europa were known to scientists as early as at the end of 1997. The moon disturbs Jupiter's magnetic field in a way that can best be explained by the presence of an electrolyte in its interior. Both Europa and—to everyone's surprise—Callisto show evidence of an induced magnetic field, the source of which can only be Jupiter. An induced field of this sort could come about if the ocean theorized for Europa consists not of pure water, but of salty water, which is a likely scenario. If Jupiter's powerful magnetic field sweeps through an ocean 10 to 100 kilometers deep, it would generate so-called cyclonic storms that would produce their own magnetic fields, which would overlay Jupiter's and have roughly the op-

The best Galileo view of Europa's cycloidal double ridges, in the Northern hemisphere. They form a repeating pattern of arcs, stretching for hundreds of kilometers.

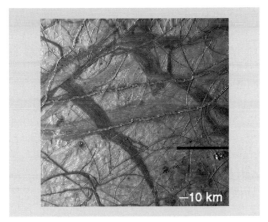

The dark structures are extensional wedge-shaped bands—they probably were initiated as cycloidal cracks.

posite charge. This induction hypothesis has not yet been proved for Europa or Callisto. Researchers may have a hard time dealing here with the complicated plasma processes in Jupiter's atmosphere. There are even reasons to believe that plasma currents around Europa cause yet another kind of magnetic field effect. Nothing has changed, in other

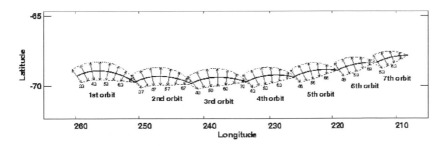

A model of cycloidal crack formation on Europa. The arrows represent the amplitude and orientation of thetensile stress, the numbers below the arrows indicate hours of the orbit. During the first orbit, cracking is initiated 33 hours after perijove; at 71 hours tension decreases to a point where the crack stops for a while. While it progresses, the orientation of the tensile stress vector changes all the time, therefore the crack changes course continuously. And on subsequent orbits, the process repeats—that is how the bizarre cycloidal chain is formed. (Adapted from Hoppa et al., *Science* 285, 1999.)

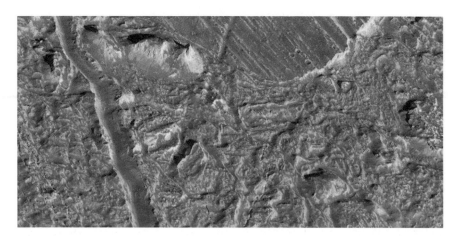

This view of the Conamara chaos region shows how the surface ice has been broken up into many separate plates, which have shifted laterally and rotated relative to each other. These plates are connected to a deeper topographical "matrix," possibly in the form of water, mud, or warm liquid ice that has risen up from underground regions. A single plate can be seen as the flat, line-covered area at the top of this picture. The image covers only 4 × 7 kilometers, with a resolution of 9 meters per pixel.

This view of the Conamara chaos region shows cliffs that formed when one edge of an ice plateau rose up. The washboard pattern in older areas is made of broken-up plates separated by material with a very irregular structure. Again, this is an extreme close-up, encompassing only 1.5 × 4 kilometers.

This very high resolution photograph of Conamara's chaos region shows an area in which the ice plates have broken apart and then shifted relative to each other. The upper part of the image is dominated by furrowed plateaus ending in cliffs over 100 meters high. Fragments that have fallen down from the cliffs can be distinguished, and a fault cuts horizontally through the middle of the image.

words—the more closely scientists examine the new improved data from a mission, the more confused the picture becomes . . . .

The breakthrough came on January 3, 2000—after the Galileo Europa Mission was over and many had presumed that the entire Galileo mission was over as well. NASA, however, had somehow scrambled together enough funds to keep the trusted spacecraft going for at least another few months and to grab the opportunity of a particularly close flyby at Europa, number E26. This was not just another opportunity to take pictures—now was the magnetometer's turn! What distinguished this Europa flyby from all that had happened before was that the moon was located far south of Jupiter's magnetic equator in a region where the radial component of the magnetospheric magnetic field points inward toward Jupiter. This pass, with a previously unexamined orientation of the external forcing field, was finally able to distinguish between an induced and a permanent magnetic dipole moment model of

Europa's internal field. And the verdict from the data was clear: If the magnetic field Galileo had seen during the previous flybys had been caused by a permanent dipole moment, it would have been oriented just the other way around than the configuration which Galileo actually encountered during the January 2000 flyby!

"The Galileo magnetometer measured changes in the magnetic field, predicting that a current-carrying outer shell, such as a planet-scale ocean, is present beneath the icy surface," the magnetometer scientists around Margaret Kivelson stated. They could finally conclude: "The evidence that Europa's field varies temporally strengthens the argument that a liquid ocean exists beneath the present-day surface"— there is just no mechanism possible by which a permanent dipole anchored in a solid core would suddenly switch its orientation just because the moon is in another position relative to Jupiter. "The case for a subsurface electrical conductor on a planetary-wide scale passed the test of the E26 flyby with flying colors," the authors write enthusiastically. "Although the electrically conducting layer need not be salty water, water is the most probable medium on Europa. Geological evidence has been interpreted as consistent with surface effects of subsurface liquid water, but the defining features could have been formed in the distant past. The magnetometer result makes it likely that liquid water persists in the present epoch." This is "certainly one of the most important discoveries of the Galileo mission," says planetary scientist Jonathan Lunine.

"A global layer of water with a composition similar to Earth seawater and a thickness greater than about 10 kilometers could explain the data," which "match the model well," elaborates David Stephenson. "This is all the more remarkable considering that the model has no adjustable parameters." And what else have we learned? "The dominant source of ions in Europa's ocean may be different from those in Earth's oceans, but they should satisfy the conductivity requirement. A

Mountain ridges on Europa. This region is mostly light colored, with darker material visible in the valleys. Part of this dark material presumably moves down along the sides of the ridges, accumulating at the bottom. The most prominent ridges are about 1 kilometer wide. The brightness of the region suggests that much of Europa's surface is covered by a layer of frost.

much thicker layer of water ice, even if it is heavily contaminated with frozen brine, cannot explain the data because the ions are relatively immobile compared with those in liquid water." Although "some exotic possibilities cannot be excluded (such as graphite or some other relatively high conductivity material, plausibly carbon-rich, intermingled within the ice but interconnected at the grain size scale)," a simple layer of water "is the most plausible explanation. A compelling demonstration of its existence or absence may be reached from gravity and altimetry data in the proposed Europa orbiter. The predicted diurnal tidal amplitude is over an order of magnitude larger for a Europa with a global ocean than for a Europa without one."

While news about Europa's intriguing interior is always a guaranteed headline, there are also new insights regarding its surface—not the morphology but the chemistry of it. Spectra from the NIMS show clearly that there is a lot of sulfuric acid mixed into the surface ice. This detection supersedes the earlier interpretation of the NIMS spectra as evidence for various salts. Laboratory experiments on icy mixtures have since been performed that make it all but certain that hydrated sulfuric acid is a major constituent of Europa's surface and can explain all the mysterious distortions seen in the water absorption bands in the NIMS spectra. Europa's surface composition varies widely, though, with some regions being pure ice and others predominantly hydrated—and everything in between. Spectral maps of Europa show furthermore that the presence of sulfuric acid correlates nicely with the darkness of the surface (for which polymerized molecules are responsible). Even some of the largest dark bands show up in the NIMS map!

While the sulfur most likely came from elsewhere in the Jovian system as ions and was implanted into the ice (where it turned to H_2SO_4 by radiation), there must also be a geological process at work that concentrates the acid in certain areas. The correlation with the dark bands makes it likely that in these locations upwelling water ice lava has freed sulfur components embedded in the crust and exposed them in a concentrated fashion. And what does the sulfur have to say about the possibility of life on Europa? "To make energy, which is essential to life, you need fuel and something to burn it," says Kenneth Nealson, a NASA astrobiologist. "Sulfur and sulfuric acid are known oxidants, or energy sources, for living things on Earth. These new findings encourage us to hunt for any possible links between the sulfur oxidants on Europa's surface, and natural fuels produced from Europa's hot interior." So, even when they discover a particularly nasty chemical compound like sulfuric acid on another world, NASA's astrobiologists will always see a silver lining . . . .

Beyond the Visible: Jupiter's Powerful Magnetosphere

It was 1958 when the U.S. made its first successful launch of an Earth satellite, a few months after the Soviet Union had inaugurated the space age by launching one of its own. It may have been Wernher von Braun's rockets that made the Explorer launches possible, but there was another scientist involved in the effort who would become at least as famous. James van Allen had installed a few small instruments on board the satellites, and the surprises were not long in coming. "With the help of simple radiation detectors on board the U.S. satellites Explorer 1 and 3, my students and I discovered a large number of energy-rich, charged particles caught in the Earth's outer magnetic field," van Allen wrote later on. "No one had expected such a finding," which Soviet scientists were quickly able to confirm. Nevertheless, the theoretical framework for the existence of such "radiation belts" had been laid down half a century earlier. In 1907, Carl Störmer had made calculations proving that electrically charged energetic particles could become permanently trapped in the field created by a magnetic dipole.

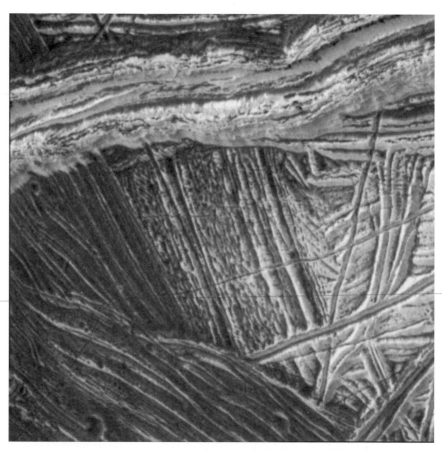

A region on Europa where pieces of the crust are moving away from each other. Visible at the lower left is part of a typical dark wedge-shaped area that has risen up to fill the cracks between the drifting plates. A linear structure can be made out running alongside the wedge, although the lines are not oriented the same way. To the right in the image is the older, brighter background crisscrossed by ridges. A large bright ridge runs east-west through the upper part of the image, cutting across both the older plains and the wedge.

The particles would be kept moving by so-called "Lorentz forces" in an arc back and forth between the poles without losing any energy.

Störmer had also shown that there was no obvious way for a particle to become trapped in such a "magnetic bottle." Once inside, however, in the simplest conceivable scenario, it could stay trapped forever. In the years after 1958, the radiation belts surrounding even so modest a planet as Earth proved to be extraordinarily complex and changeable. The term *magnetosphere* was quickly introduced to describe the entire region of such "trapped radiation." The entire magnetosphere is very dynamic, but change is constant especially in its outermost regions, where it is affected in particular by changes in the solar wind. Forty years after the Explorers, we are still launching probes to investigate individual aspects of the Earth's magnetosphere, more probes than all those under way to other planets combined. The Equator-S, a German satellite launched on December 3, 1997, is investigating yet another unexplored region of our magnetosphere.

Our understanding has also been advanced by explorations of the magnetospheres around other planets in the Solar System, most especially the gigantic one that surrounds Jupiter. But astronomers' exploration of Jupiter's magnetosphere did not begin in 1973 with the arrival of the first interplanetary probes. Two decades earlier, radio telescopes had picked up signs of some very energetic phenomena going on around Jupiter, and sporadic bursts of radio emissions coming from the Jovian system had been discovered in 1955. At a frequency of 22.2 megahertz, these were referred to as decametric waves, and they were definitely "nonthermal" in nature, which identified them as something other than the heat normally given off by all astronomical bodies. Based on the decametric waves, scientists would soon be able to pin down Jupiter's "true" rotation period, independent of the confusing cloud currents. And in 1958, another form of nonthermal radiation was discovered at much higher radio frequency (300 megahertz versus 3 gigahertz), with wavelengths in the decimeters (10 centimeters and more).

This radiation—its power measured in the billions of watts—was practically constant, and it came from a ring-shaped region surrounding Jupiter. In 1959, scientists correctly interpreted it to be synchrotron radiation from trapped electrons in a radiation belt around Jupiter. The gas giant had a radiation belt not unlike the one that had been discovered just a year earlier around Earth. These were already generally known as van Allen belts. Actual confirmation of Jupiter's radiation belt finally came with the Pioneers 10 and 11. As they flew dangerously close to the planet, their instruments detected precisely the distribution of electrons that scientists had already identified as the decimeter radiation. The success proved that the physics of synchrotron radiation by electrons in magnetic fields was well understood, and that exploring decimeter radiation would be a good way to study the magnetosphere in detail.

A Magnetosphere Like No Other Seen Before

Modern radio telescopes make it possible to resolve spatial images of Jupiter in the "light" of synchrotron radiation, using them to explore the magnetosphere. Jupiter's resembles the magnetic field around our own planet, despite containing perhaps a million times the number of charged particles per unit of volume and being about 1,200 times bigger. On the side Jupiter turns to the Sun, the magnetosphere extends out into space for a distance of 25 to 50 Jupiter diameters, depending on the force of the solar wind pressing back on it. On Jupiter's far side, the wind pushes it back into a gigantic "magnetotail." If it could be seen, it would appear from Earth to be several times larger than the full Moon.

The magnetotail extends for hundreds of millions of kilometers in the direction away from the Sun. Aside from the tails of very rare giant comets, it is the biggest single structure in the entire Solar System.

Such enormous dimensions stem, on the one hand, from the fact that the solar wind is only 1/25 as strong as it is here by the time it reaches Jupiter. But the main determining factor is Jupiter's magnetic moment, which is 19,000 times Earth's! Because Jupiter is so much bigger than Earth, however, the magnetic field at the upper edge of the clouds is only about 12 times more powerful than it is on Earth's surface. The tilt of the field relative to the rotational axis, incidentally, is 11°, compared to 12° on Earth.

Jupiter's magnetosphere is shaped in essence like an enormous disk, which is the result of two different processes. First, the mass of

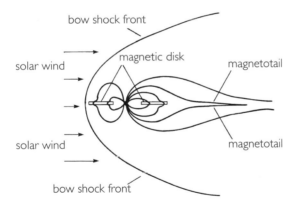

The large-scale structure of the Jovian magnetosphere, the largest long-lasting structure in the Solar System. A bow-shaped shock wave arises at the point where the solar wind encounters the magnetosphere, similar to what happens in front of a ship moving through water. Particles in the solar wind stream around the magnetic field, which trails behind the planet like a long tail. Because the intensity of the solar wind varies, it pushes with greater or lesser force against the magnetic field at different times, causing the distance between the shock wave and the planet to vary between 25 and 50 Jupiter diameters. Outward pressures in the equatorial region, together with the planet's rapid rotation, force the magnetosphere into the shape of a thin disk. A current flows in a ring around Jupiter inside this disk.

"Killer Electrons" from Jupiter

The bombardment of Earth by extremely high-energy electrons comes from the two giants of the Solar System: the Sun and Jupiter. We owe this astonishing finding mainly to NASA's SAMPEX satellite. Along with a number of other Earth-orbit satellites involved in the International Solar Terrestrial Physics Program, SAMPEX has now put together a complete image of Earth's particle environment—which can be a dangerous place.

When geophysicists speak of "killer electrons," they mean electrons with extremely high energy, and they can pose a danger to satellite electronics. Several failures or outright losses of satellites already have been traced to electron bombardment, making it one of the least hospitable aspects of the "weather" in space.

The Sun is responsible for many killer electrons. Disturbances in the solar magnetic field operate like particle accelerators, propelling the electrons through space at velocities far in excess of that of the normal solar wind. It takes the wind two to three days to travel the 150 million kilometers to Earth (see the text box on page 250)—killer electrons cover the distance in 20 minutes, traveling at nearly the speed of light!

The Sun naturally dominates the energy flows around Earth. Nevertheless, according to the new findings, whenever solar activity temporarily subsides, Jupiter steps in to take over the "job." Jupiter's magnetosphere is also an effective particle accelerator, capable of raising electrons to yet higher energies than the Sun. SAMPEX has now identified electrons of precisely that sort coming from Jupiter during periods of reduced solar activity. These findings could improve the long-term ability of scientists to predict the weather in space, which is a particular concern of the operators of communications satellites. Everything in the Solar System truly is related to everything else—and the sphere of influence of the largest planet reaches all the way to us. As one SAMPEX researcher summarized the new findings, "in many ways, Earth is like a cork bobbing up and down on the currents flowing back and forth between the Solar System's giants."

the low-energy plasma trapped inside it is so immense that it causes the magnetic field to expand like an air balloon. And because the field is weakest over the equator, the magnetosphere swells out most at that point. Second, because the plasma has an electrical charge, it has to move along with the magnetic field, which means keeping up with Jupiter's rapid rotational period (spinning once around on its axis

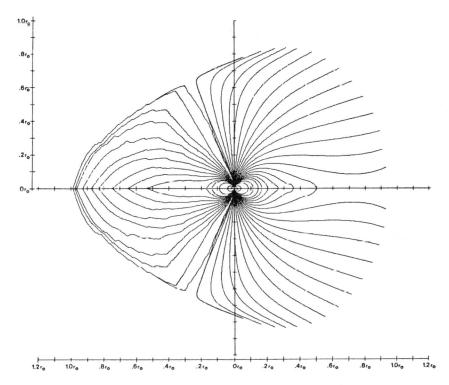

Large-scale structure of Jupiter's magnetosphere, depicted here in a mathematical model from 1980. The lines indicate where the magnetic field has the same strength. The axes are in units of solar radii.

every 9 hours, 55 minutes, and 29.7 seconds). These particles remain completely in the grip of the magnetic field at least as far as ten Jupiter diameters away. And centrifugal force joins the other forces pushing outward, making Jupiter's magnetosphere into more of a "magnetodisk," in fact—but that term, once proposed, failed to stick.

The magnetosphere can also function as a gigantic particle accelerator, sending ions racing at speeds in the tens of thousands of kilometers per second. A few of these particles are hurled into space, to collide with the Voyager spacecraft, for example, as much as 50 million kilometers away from the planet. A Jupiter particle will occasionally blunder all the way to Earth! Remarkably, temperatures in the magnetosphere range from what is easily tolerable for a spacecraft in the outer regions (where density falls to extremely low levels of one particle per 100 cubic centimeters) to hundreds of millions of degrees in the interior. Here there is enough energy in this extremely hot plasma for it to be primarily responsible for fending off the solar wind on the edge of the magnetosphere. The Pioneer mission had already established the fundamental shape and physics of Jupiter's magnetosphere, but understanding the chemistry, the way the plasma particles are composed, had to wait for the improved plasma analyzers on board the Voyager probes.

The plasma comes mainly from sulfur dioxide, hydrogen sulfide, and other gases that are released by Io's volcanos, leading to the emergence of a special plasma torus that exists nowhere else in the Solar System. On the other hand, Jupiter's moons also play a critical role in the absorption of particles from the radiation belts, because some of them travel right through the magnetosphere, swallowing up everything they come across. In the process, the moons also give up molecules from their own ice crusts, which are released through the impact of the energetic particles. Projections regarding Europa's ice crust (made possible by data from the particle detector EPD) show that every million years a layer of ice about 80 centimeters thick is chipped off in this process that chemists call "ion sputtering," only to settle back down somewhere else on the moon's surface. About 20 centimeters every million years are lost into space, but that loss may be compensated by a comparable rate of ice formation from below.

The escaped particles become part of the magnetosphere, in a process that especially affects Io, the moon that is most "bound" to Jupiter. In a way that remains very poorly understood, Io seems to be principally involved in the outbursts of decametric radio waves mentioned earlier. They are especially frequent whenever Io, Jupiter, and the Earth are positioned in a specific configuration with one another.

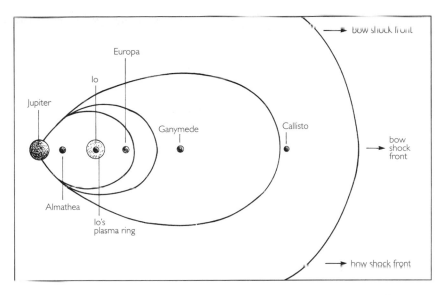

The four Galilean moons are all embedded in Jupiter's magnetosphere. On each revolution, as it moves between Jupiter and the Sun, Callisto comes close to the bow-shaped shock wave. All of the moons are under constant bombardment by energetic particles trapped in the magnetosphere, a situation that affects the surface of the moons and also keeps the magnetosphere charged with new particles.

And they account for a much greater share of Jupiter's overall radio emission than the more constant decimeter radiation.

Instabilities in Jupiter's plasma environment may be responsible for the decameter outbursts and for radio emissions at much longer "kilometric" wavelengths. The latter were first discovered on the Voyager fly-bys in April 1978. Unlike in synchrotron radiation, the reciprocal effects of electrons on each other play a central role here, not just their individual motion. Since there is practically no limit to how complicated the physics of this process can become, let us return to the general characteristics of the fields and the particles that make them up. Our understanding of the structure of Jupiter's ionosphere is ultimately limited. High levels of electron density can be identified in several different states a few

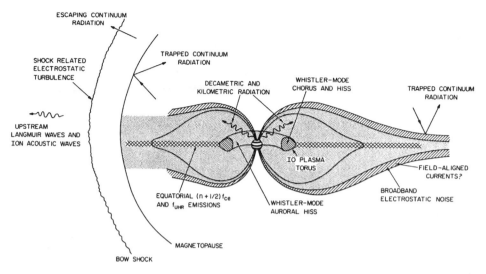

This diagram, based on readings by the Voyager probes, shows sources of radio waves everywhere in the magnetosphere.

thousand kilometers above the clouds. Galileo used radio science to investigate them, shortly before it disappeared around the other side of the planet on December 8, 1995. The real drama, however, takes place in the surrounding magnetosphere.

The energy of particles in the magnetosphere around Jupiter typically reaches levels about ten times those found inside Earth's magnetosphere, and the intensity of the radiation trapped in Jupiter's magnetosphere is several magnitudes higher. Also of note is the famous "flux tube" connecting Io and Jupiter, which carries a current of 5 megaamperes. When Galileo approached Io in 1995, its Energetic Particle Detector (EPD) discovered something new in this connection. Flowing in both directions, exactly parallel to the magnetic field, is intense electron radiation, which constitutes a direct proof that Io and its flux tube actually amount to a highly efficient particle accelerator. This electron radiation, on the one hand, may be part of the long-sought

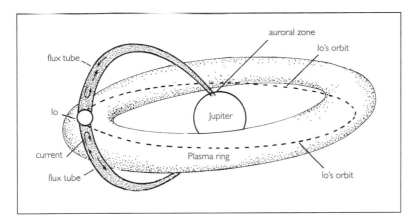

Io's flux tube and plasma ring, showing the two fundamental reciprocal effects that take place between Io and Jupiter's magnetosphere. The central axis of the plasma ring coincides roughly with Io's orbit, and the ring's thickness is about equal to the diameter of Jupiter. Energetic sulfuric-acid and oxygen ions course through the ring at a temperature of roughly 100,000° Celsius. Since the axis of the magnetic field is tilted in relation to Jupiter's rotational axis, Io moves in and out of the ring during each revolution around the planet.

mechanism that produces the decametric radio radiation mentioned earlier. But, on the other hand, it contains so much energy that it could explain the auroral lights that occur where the tube connects with Jupiter's poles, causing the gases there to glow in a number of wavelengths, including the infrared. The "footprint," where the tube makes contact with Jupiter's atmosphere, also provided the Hubble Space Telescope with a prime target for exploration in the ultraviolet.

Jupiter's Polar Lights

Polar lights, or auroras, are phenomena that tend to go with magnetospheres. The charged particles trapped in the magnetosphere are more prone to escape in the polar regions than anywhere else, coming into

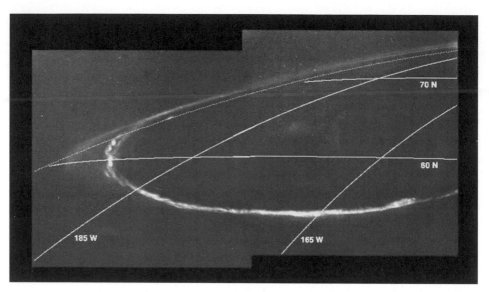

Jupiter's aurora on the nighttime side of the planet, in a photograph taken in visible light on November 5, 1977. The coordinate grid has been superimposed.

contact with the atmosphere and causing certain kinds of gas to glow. They also set off chemical reactions that darken the clouds near Jupiter's poles in most of the wavelengths. Voyager 1 had already discovered the ultraviolet emissions that characterize auroras on Jupiter's daytime side, and moving around the planet, it registered a bright aurora over the planet's north pole on the dark side. The job of monitoring Jupiter's polar lights was soon taken over by the International Ultraviolet Explorer (IUE), the European Space Agency's satellite observatory, without, however, the capacity to resolve the lights spatially. Since the launch of the Hubble Space Telescope in 1990, astronomers have been able to get extremely sharp images of Jupiter's polar lights in the ultraviolet. And, assuming they manage to claim some of its hopelessly over-booked observation time, they will be able to track changes in the lights over time.

It turns out, for example, that Jupiter's auroral ovals rotate along with Jupiter and its powerful magnetic field, whereas on Earth they maintain a constant position relative to the Sun (more precisely, to the solar wind). The difference has to do with the fact that the energy for the polar lights on Earth comes ultimately from the solar winds themselves. On Jupiter, the lights draw energy from the planet's own rotation. The ovals are always closer to the poles than the footprint of the flux tube is, which provides scientists with information about the origin of the charged particles raining from the magnetosphere onto the atmosphere. From projections made of the lines of force in the magnetic field, the source of the auroral particles appears to be farther away from Jupiter than Io. It is at about the magnetosphere's midpoint, about six Jupiter di ameters away. It has always been a puzzle how the particles, upon escap- ing the magnetosphere, could possibly get started on a course that ended in the atmosphere. During its first Ganymede encounter, Galileo was ac- cidental witness to the temporary deformation of Jupiter's magnetic field, which demonstrated the anticipated effect. In principle, it has to do with the magnetosphere becoming "too full," whereupon a "bubble" forms and separates off. And then the whole process begins anew.

The auroras, in other words, seemed to be involved in the complex reciprocal relationship existing among Io, the solar wind, and Jupiter's magnetic field. The possibility that this might be true called for coordi- nated observations with all available instruments. The IUE satellite made its principal contributions to the International Jupiter Aurora Watch in the summer of 1996. Better than any other instrument, it was able to register in detail the surprisingly rapid variations taking place in auroral activity, while Hubble (and Galileo, for that matter) could gather such data only intermittently. Both the appearance and the intensity of Jupiter's auroras varied not only in the short term, but over

Jupiter Aurora
Hubble Space Telescope · WFPC2

PRC96-32 · ST ScI OPO · October 17, 1996 · J. Clarke (University of Michigan) and NASA

Jupiter's aurora and its relation to Io, shown in images from the Hubble Space Telescope taken from May 1994 to September 1995. Sketched onto the visible-light photograph, above left, is the flux tube between the moon and the planet; to the right of that is the view in the ultraviolet 15 minutes later. Visible here are the auroral ovals—and Io's so-called footprint, where particles from the flux tube collide with Jupiter's upper atmosphere, causing the hydrogen gas to glow. Io has "disappeared" in this image, because it is very dark in the ultraviolet.

The two UV photographs (bottom) show how the view of Jupiter's auroras changes with the rotation of the planet. This is caused by the 10° to 15° tilt of the magnetic field relative to the planet's rotational axis. (Photographs: Clarke, Ballester, Trauger, Evans, & NASA.)

the course of years. Scientists determined that the variable volcanic activity on Io was responsible for the long-term effect. This activity had an immediate influence on the abundance of ions in the magnetosphere. Galileo's EPD instrument could follow that variation directly.

Jupiter's polar lights shine most brightly in the Lyman-alpha emission line of hydrogen at 122 nanometers. During the International Jupiter Aurora Watch it had been possible to compare changes in the intensity of L-alpha radiation with conditions in the solar wind, which was likewise being recorded "live" by an international team of specialists. Although there was little solar activity, the brightness of the Jovian aurora varied by a factor of two to three. Once scientists fully understand this relationship, they may even be able to use Jupiter's auroral ultraviolet radiation as a solar wind "detector" in the outer Solar System.

And what did Galileo see of the polar lights? On April 2, 1997, the camera had a look in five different wavelengths at once. It succeeded in taking the first pictures showing the aurora shining in visible light against the dark background of the planet. The brightness values associated with the various filters offer information about the charged particles in Jupiter's upper atmosphere which are causing the lights. For example, if the radiation were coming exclusively from hydrogen, it would appear through a red filter to be much brighter than it does. Pictures from Galileo also gave us our first information about the altitude of the aurora in Jupiter's sky, about how thick the auroral oval is, and about the whole structure in detail.

Jupiter's Polar Lights: A Case for Hubble

Not even an all-round spacecraft like Galileo can be optimally equipped for every purpose. In one area of Jupiter research, an Earth-based telescope—more precisely, a telescope in Earth orbit—was able to make a major contribution of its own. Jupiter's polar lights are bright primarily in the ultraviolet, a range of wavelengths to which Galileo's CCD (charge-coupled device) camera is barely sensitive. The Hubble Space Telescope, however, carries a number of instruments that are better suited to such short wavelengths—three, in fact, since the retrofits made on the telescope in early 1997. Just after it was launched, Hubble took some astonishingly clear photographs of Jupiter's northern lights. They were the first pictures ever of the phenomenon, which, although discovered in 1980, had never been spatially resolved. Jupiter is almost black in the ultraviolet, and by observing the ultraviolet emissions of the hydrogen molecule, Hubble was able to distinguish the auroral lights.

For the first time scientists could see that the lights—much like the polar lights on Earth—are confined to a rather precisely defined oval. Improving Hubble's optics also drastically increased both the quality of the Jupiter photographs and the details of the ovals it became possible to make out. Images from 1994 and 1995 captured the ovals around both poles, showing how they are constantly changing. Diffuse lights inside the ovals, toward the magnetic poles, could also be seen, making it apparent for the first time that Jupiter's northern and southern polar lights, like Earth's, are mirror images of each other. And that was not anticipated. Whereas the auroras on Earth are fed by the solar wind, the charged particles causing Jupiter's atmosphere to glow come from its own magnetosphere. Theorists had expected that the complicated distribution of the particles in the atmosphere would cause the shape of the northern and southern lights to be completely different.

Since they are so closely bound up with Jupiter's magnetosphere and magnetic field, the ovals rotate approximately with the planet. This is also different on Earth, where the Sun determines the orientation of the auroras. The Hubble pictures also showed Io's "footprint," where the moon's flux tube intersects with Jupiter's atmosphere. The particles streaming through here cause an area to glow which is 1,000 by 2,000 kilometers in size. Io's footprint would be a dramatic sight for an observer positioned at the level of Jupiter's clouds. It follows along with Io, of course, racing around the planet at 5 kilometers per second. The whole sky about 400 kilometers above the observer would suddenly be aglow, and then the light would slowly fade. Color plate XXIII shows the best view humans have had so far of Jupiter's polar lights.

Io the Miracle Moon: Endless Volcanic Activity

Io, 3,640 kilometers in diameter, is the only object of its kind in the Solar System. Of all the Galilean moons, Io orbits closest to Jupiter, and it has landscape formations like no other moon, or for that matter, planet. There are no impact craters marring it singularly reddish-yellow surface, contrary to what is observed on other solid planetary bodies. Instead, Io has a multitude of volcanic structures, looking exactly as we might imagine. Io is much less red in color than it appeared in the first Voyager photographs; instead it is more a variation among olive green, yellow, and orange. This coloration comes primarily from the chemical element sulfur. Its different molecular forms exhibit a range of yellow and red tones very much like what we observe on Io. The conclusion, however, that Io's volcanos run entirely on sulfur proved false. Normal silicate-based lava with some sulfur content can also take on a reddish color. In any case, Io represents a distant paradise for volcano specialists. There are plenty of calderas, both with and without visible flow structures.

A caldera (the Spanish word for "caldron," first applied to the central volcano on La Palma in the Canary Islands) is a ring-shaped crater-like basin left over from the collapse of one volcano after a new volcano has erupted in the middle. As many as 200 calderas with diameters of more than 20 kilometers are responsible for Io's frankly pockmarked appearance. Earth, which is three and a half times as big, has barely more than a dozen of such volcanos, whereas on Io they make up a full 5 percent of the surface. To optical cameras the calderas appear pitch black. They reflect less than 5 percent of the sunlight. Unlike many volcanos on Earth and Mars, on Io tall shield volcanos do not form at the eruption sites. Obviously, the lava on Io is much thinner, so that instead of piling up, it flows out in all directions, forming pools typically 100 kilometers across. The successive layers are sometimes brighter, sometimes darker than the background. Nevertheless, of all the Galilean moons, Io has the most dramatic topography. The existence of mountains on Io proves that its surface, at least in places, is relatively solid. Where the mountains actually came from remains unclear, but certainly their origin is not volcanic.

With the help of Galileo's images German photogrammetrists could actually determine the precise shape of Io. The moon is not really a sphere but a three-axial ellipsoid, with the longest axis pointing toward Jupiter. The shortest axis—the one in the direction of Io's orbital motion—is 15 kilometers shorter. And there is more: The tidal action by the other moons deforms Io measurably and periodically. This, of course, is the reason for its stupendous internal heat and all the volcanic activity. In early 1999, it was nice to see these physics at work in actual measurements which had been predicted exactly 20 years earlier.

The most spectacular events on Io are without doubt the *active* volcanos, with their towering plumes. For Voyager camera operator and Galileo mission head scientist Torrence Johnson, they are among "the most impressive and beautiful sights in the Solar System." The weakness of Io's gravity allows the plumes to spurt to heights of as much as 280 kilometers, where they form an umbrella shape reaching all the way back down to the surface. Most features suggest an eruption speed out of the craters of 0.6 to 1.0 kilometers per second. The particles making up the plume follow ballistic trajectories, descending in a cross-shaped zone with a diameter of 1,400 kilometers. The flow speed is extreme, even compared to explosive eruptions on Earth, where a flow of just 100 meters per second is fast. Strictly speaking, the Io volcanos are geysers, which makes them conform better to our scale on Earth. If it were possible to transplant the famous "Old Faithful" geyser in Yellowstone National Park to Io, its plume would shoot to a height of 35 kilometers. Geysers on Earth are driven by rapid phase changes in water—from water to vapor, for example.

1 km (0.6 mile)

The margin of the lava flow field associated with the Prometheus volcanic plume on Jupiter's moon Io is seen in this image, taken by Galileo on February 22, 2000. The image has a resolution of 12 meters per pixel. The dark lava has margins similar to those formed by fluid lava flows on Earth. Because this entire area is under the active plume of Prometheus, which is constantly raining bright material, Galileo scientists interpret the darkest flows as being the most recent. They are not yet covered by bright plume fallout, perhaps because they are too warm for bright gas rich in sulfur dioxide to condense. The older plains (upper right) are covered by ridges with an east–west trend. These ridges may have formed by the folding of a surface layer or by deposition or erosion. Bright streaks across the ridged plains emanate from the lava flow margins, perhaps where the hot lava vaporizes sulfur dioxide.

This mosaic of images collected by Galileo on November 25, 1999, shows a fountain of lava spewing above the surface of Jupiter's moon Io. The active lava was hot enough to cause what the camera team describes as "bleeding" in Galileo's camera; this occurs when the camera's detector is so overloaded by the brightness of the target that electrons spill down across the detector, showing up as a white blur in the image. Most of the hot material is distributed along a wavy line which is interpreted to be hot lava shooting more than 1.5 kilometers high out of a long crack, or fissure, on the surface. There also appear to be additional hot areas below this line, suggesting that hot lava is flowing away from the fissure. Initial estimates of the lava temperature indicate that it is well above 1,000 Kelvin and might even be hotter than 1,600 Kelvin.

On Io there is no water. It was probably eliminated by the high temperatures, although it is also possible that the moon was formed without any water. Some other liquid must be driving the volcanism. Judging from the high sulfur content on the surface, that element may well play a critical role. Both sulfur and sulfur dioxide are likely to be present in liquid form just a few kilometers deeper inside Io, at least according to current models. Nevertheless, individual eruptions on Io can reach temperatures of up to 1,700° Celsius, which is much too hot to be driven purely by sulfur. Sulfur boils before it gets nearly that hot.

Io's volcanism is much more likely being driven by silicates, with new surface formations taking shape with such speed on Io that impact craters are quickly erased. Each year, on average, it has to produce a millimeter or more of fresh surface, so that over the history of the Solar System, Io's crust and mantle have been reformed many times. No part of Io's surface as we see it today can be more than a million years old. The level of volcanism required here would have been inconceivable for a body as small as Io—if it had not just been discovered how Jupiter and the other large moons are constantly pummeling it with tidal forces. Voyager 1 alone sighted eight different volcanos going off at once on Io, and Galileo witnessed a comparable level of activity. It would take Earth a century to account for as much.

Hot Spots Everywhere

In both 1996 and 1997 Galileo was able to follow innumerable volcanic eruptions from a distance. Some of the sites were the same as the ones the Voyager probes passed by in 1979, but changes were also evident. In some places, new lava flows had been created, and in others, volcanos had wandered as much as 60 kilometers, leaving a dark trail behind them. Already during the first Ganymede encounter, for example, Galileo had discovered new streams and reservoirs of lava around Ra Patera, a volcanic peak already known as a distinct brightening on photographs of Io made by the Hubble Space Telescope in 1995. There was no hot spot to go with this glow, however, which, in combination with the yellowish color of recent surface strata, argued for typical sulfur volcanism. And yet, subtle changes were evident in the landscape around the known hot spots Loki and Kanehikili, where silicate vulcanism is without doubt under way. Among the first impressions made by the Galileo data was the intricate complexity of Io's volcanism. In other places, during the 18 years since the Voyager photographs, only the color of the volcanic deposits had changed. For scientists, all this was something like time-lapse geochemistry.

The most spectacular change in the first two years of Galileo's observations was the creation of a volcanic deposit the size of Arizona. Stretching 400 kilometers from side to side, it was produced by the volcanic center Pillan Patera as if on schedule for the Galileo mission from April through September 1997. In June, both Galileo and Hubble

Figure 2

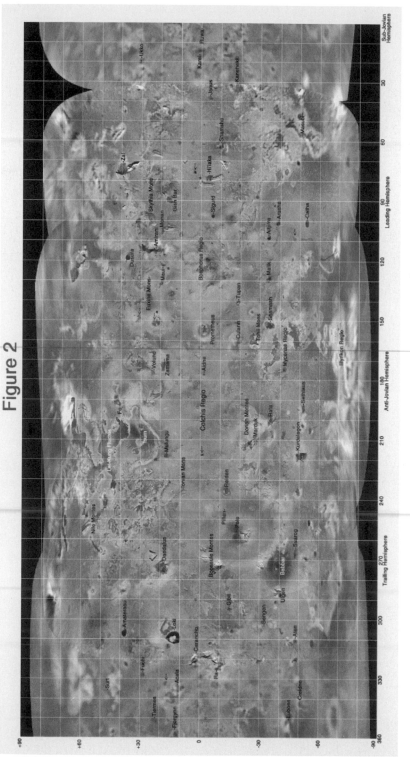

This is the best complete map of Io we have: a global mosaic combined from the highest resolution Galileo images. The individual pictures were obtained at low sun illumination angles that emphasize topographic shadows over the course of several orbits. Named features are marked while several active but as yet unnamed volcanoes are indicated by arrows.

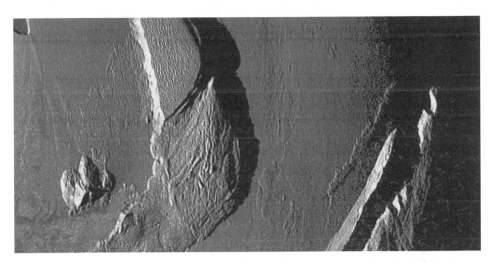

This image, taken by Galileo during its close Io flyby on November 25, 1999, shows some of the curious mountains found there. The Sun is illuminating the scene from the left, and because it is setting, it exaggerates the shadows cast by the mountains. By measuring the lengths of these shadows, Galileo scientists can estimate the height of the mountains. The mountain just left of the middle of the picture is 4 kilometers high, and the small peak to the lower left is 1.6 kilometers high. These mountains, like others imaged during a previous Galileo flyby of Io in October 1999, seem to be in the process of collapsing. Huge landslides have left piles of debris at the bases of the mountains. The ridges that parallel their margins are also indicative of material moving down the mountainsides due to gravity.

had sighted a plume spraying to a height of 120 kilometers, and infrared astronomers had noted an extraordinary hot spot. The gray color of the surrounding deposits is unusual, however. In most other places they are white, yellow, or red; these new ones probably consist primarily of silicates. A more detailed analysis could determine the exact minerals that might be involved. There is no disputing that Io is made up mainly of silicates, but its extreme volcanism may have led to effects not found on Earth. Galileo was also able to register volcanic activity on several flybys. For NIMS, Io's surface is practically covered with hot spots, and in Jupiter's shadow even the optical camera was able to capture the hottest spots. Before, scarcely anyone had thought it possible that the camera, sensitive to wavelengths measured in micrometers, would be able to register the hot spots at all.

Hot spots reaching very high temperatures exist in abundance on Io, as scientists determined while analyzing the pictures of the shadowed moon. They identified about 20 different areas at least 100 by 100 meters in expanse with temperatures of 700° Celsius or more—another strong indication of silicate volcanism, because in a vacuum, sulfur boils at just 200° Celsius. NIMS images showed as many as a dozen different hot spots going simultaneously at any given time, with temperatures between −63° and +555° Celsius (whereas the prevailing surface temperature is only about −170° Celsius). An especially bright hot spot is Pele, for which Galileo estimated a temperature of about 430° Celsius. At least 12 of the 41 active centers spotted by NIMS are hotter than 1,200°C. The temperature record on Io so far stands at a whopping 1,700°C, measured at the Pillan Patera volcano. Only magnesium-rich silicates in the lava can explain temperatures this high. It makes current Io quite similar to the planets Earth and possibly also Venus and Mars in their youth. NIMS discovered a total of 18 new hot spots and registered 12 others that were already known. Most hot spots seemed to be "turned on" all the time, and 10 of the 30 spots were active during the Voyager visits. The hot spots therefore are evidently not distributed at random. They seem to form "rings of fire" surrounding the two bright regions Bosphorus Regio and Colchis Regio, which are rich in sulfur dioxide. Near the plume of Ra Patera and in other places on the edge of the moon darkened by Jupiter in the background, the camera picked up one more phenomenon, a faint glow, presumably of an aurora. The transparent dome of light, which seemed to reach all the way down to the surface, was hardly even 20 kilometers thick. Long exposures of the moon revealed more and more mysterious emissions from Io's tenuous atmosphere, and that even in three different colors. There were comparably bright blue glows coming from major volcanic plumes. Apparently electrons hitting sulfur dioxide molecules in the volcanic gasses made them shine. Then there were weaker red emissions along Io's limb and brightest on the pole close to the plasma torus. And there was a faint green glow from Io's night side. The evolution of these aurora-like features with time was also remarkable: most diffuse light faded while Io was in Jupiter's shadow—but the localized blue glows brightened instead. A lot of modelling remains to be done here.

A Massive Iron Core

Critical information about the nature of Io's interior comes from readings taken during Galileo's only close encounter so far, when it raced by on December 7, 1995, at an altitude of 899 kilometers. By analyzing the effect that Io's gravitational field exerted on Galileo's orbit, scientists would propose an inner iron core of considerable proportions for Io, extending to a radius of 900 kilometers. This was the first time that an iron core had been discovered in a moon. It must have originated at a time when Io's core was fluid, when the heavy and light chemical elements could still separate from each other. That could have been when the moon was formed, or it could have taken place later, for Io, like no other Jovian moon, is subject to tidal forces that could very well be responsible for heating up the moon. They remain the driving force of the volcanism dominating Io's surface even today. Further data analysis of Galileo's orbit finally led in the summer of 1996 to the "inescapable conclusion" that Io possesses "a large metallic core." The chemistry, unfortunately, remains unknown. Is the core made of a mix of iron and iron sulfide? If so, the core alone would account for about 20 percent of Io's mass, extending to 52 percent of the moon's radius (which is 1,821 kilometers). If it consists entirely of iron, then it makes up about 10 percent of Io's mass and 36 percent of the radius. Even then it would be

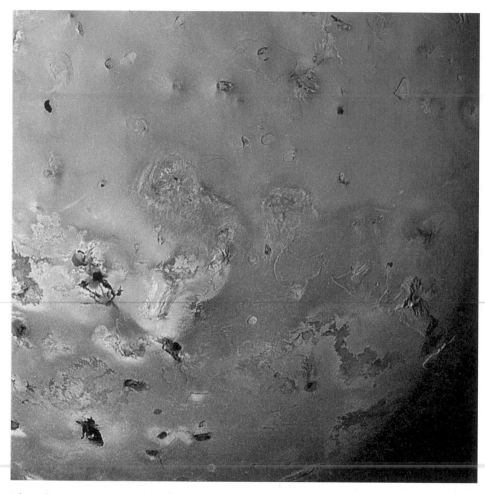

This photograph, one of the best we have of Io, was taken by Galileo during the primary mission and shows a multitude of landscapes. This panorama encompasses an area of 2,000 kilometers.

larger than our own Moon's (not yet definitively proved) iron core, which takes up at most 20 percent of the radius.

Although the rate of data transmission during Galileo's visit to Io was limited, making some images impossible, the level was adequate to convey detailed information about the surrounding magnetic field. Io produces an actual hole in Jupiter's magnetic field, causing its strength to fall off by 10 percent. Scientists initially interpreted this reading to mean that Io has its own magnetic field that partially cancels out Jupiter's. A closer look at the data, however, suggested the possibility of a different, more complex interpretation having to do with plasma physics and including a potential role for all of Io's volcanism. This explanation involves a thick ionosphere created by Io's volcanic gases. To account for the necessary effects, some plasma researchers have already sketched out complicated scenarios with gigantic invisible "stealth plumes" pouring out of the volcanos. After two years of vigorous discussion at several conferences, however, it seems that the first interpretation will prevail. Io probably has a weak magnetic field of its own which would also give it an iron core to act as a dynamo. The Galileo magnometer team concluded in October 1997 that plasma effects could account for at most 30 percent of the diminution of Jupiter's magnetic field observed in Io's vicinity, and that there was no way to avoid the assumption of a source of magnetism located inside Io. The last word may not have been spoken, however, because Galileo was to make two more close Io flybys at the end of 1999. If the orbiter holds out until then, tolerating the extreme radiation levels to which it will be exposed, our understanding of the most active solid body in the Solar System will make another leap forward.

Progress Report: 300 Active Volcanoes on Io

There may be no fewer than 300 active volcanoes on Io, a new extrapolation from Galileo observations shows. During a close flyby on February 22, 2000, NIMS detected 14 volcanoes in a region where previously only four had been known—and that region covered just five percent of Io's surface. Before that only 81 active volcanoes had been counted on the whole moon. There were also remarkable changes in the months between the three Io flybys in October and November 1999 and February 2000. Some of the smaller and fainter volcanoes appear to turn on and off, changing from hot and glowing to cool and dim within a few weeks. The larger and brighter volcanoes, however, tend to remain active for years or even decades, based on previous observations by Galileo and the Voyager visits in 1979. The February 2000 flyby yielded more images with higher resolution than previous flybys: they show unprecedented views of small surface areas that give new clues about the volcanic terrain but also reveal landforms that are perplexing. In one region, for example, there are thin, alternating bright and dark layers of unknown origin.

The continuing monitoring of Io's volcanoes by Galileo's instruments as well as from space has made it clear that many of the lava flows are surprisingly hot—over 1,200 and possibly as much as 1,300° Celsius, with a few areas even reaching 1,500° Celsius. Such high temperatures imply that the lava flows are composed of rock that formed by a very large amount of melting of Io's mantle—and this reawakened an old hypothesis that suggests that the interior of Io is a partially-molten mush of crystals and magma. That idea, which had fallen out of favor for a decade or two, explains high-temperature hot spots, mountains, calderas and volcanic plains—and if correct, Io would give us an opportunity to study processes that operate in huge, global magma systems, which may have played a role in the young Earth.

Even More: A Ring, Mini-Moons, and Dust Streams

Jupiter's ring had already been discovered in 1974. When Pioneer 11 was crossing through Jupiter's magnetosphere, coming to within 1.6 Jupiter radiuses of the center of the planet, the detection rate of energetic particles fell off whenever the probe came into the vicinity of a known moon. The moons operate as negative sources of charged particles, quite effectively sweeping them out of the magnetosphere. The detection rate also declined between 1.7 and 1.8 Jupiter radii, an area no moon was known to occupy. Some of the Pioneer scientists postulated the existence of a previously undiscovered moon or perhaps an inconspicuous ring system. Still other explanations were advanced for what had happened to the particles in this area.

Since none of the efforts to identify a ring around Jupiter from ground-based telescopes had succeeded, most astronomers remained skeptical—but, project scientists thought, why not have Voyager 1 take a look? It very quickly discovered a thin disk of matter, stretching inward toward the planet from 1.81 Jupiter radii away. Voyager 2 was then able to distinguish three main segments of the disk. Suddenly, at 53,000 kilometers above the clouds (1.81 Jupiter radii), the main segment began, extending inward 6,000 kilometers toward the planet to 1.72 Jupiter radii. The density of the matter making up the main segment, however, is extraordinarily low, so that only a tiny fraction of the space is occupied. Compared with the rings around Saturn and Uranus, Jupiter's ring appeared to be decidedly uniform. The outer 800 kilometers are only somewhat brighter than the inner region.

Jupiter's ring. Galileo was able for the first time to distinguish clear radial structures, i.e., thicker and thinner bands, for which the individual moons may be responsible. The sharply defined outer edge of the ring's main band was a surprise. Astronomers had expected the tiny particles making up the ring to be thoroughly mixed up by Jupiter's magnetosphere.

The other components of the meager ring system revealed by the Voyager pictures included an even thinner continuation of the main ring toward Jupiter and an equally tenuous halo, a lens-shaped disk made out of dust. As much as 10,000 kilometers thick in some places, the halo encloses the main disk, which is presumed to attain only a few kilometers in thickness.

Back lit, the rings appear 20 times brighter than they do in reflected light. Most of the particles making up the rings measure only a few microns in diameter. For comparison, Saturn's rings consist predominantly of boulders a few meters across. The tiny objects forming Jupiter's rings are very short-lived phenomena. They are extremely vulnerable to radiation and electromagnetic processes, which convey an electrical charge to individual particles, causing them to spin out of orbit and crash into Jupiter. In other words, there must be a constant source of new particles, which cannot be seen. In all likelihood, this source is stone boulders that are embedded in the rings but not resolved by the camera. These "moonlets" are being constantly bombarded by micrometeorites from interplanetary space and by particles spewed out of Io's volcanos. The visible rings are made of dust particles that are split off from the boulders in the collisions.

Adrastea and Metis, bodies 20 and 40 kilometers in diameter, respectively, were discovered by Voyager 2 very near the rings, and they may also play a critical role in replenishing them. They represent the tip of the iceberg, however—the only moonlets that can be seen directly. Scientists quickly sketched out the basic scenario. A considerable number of moonlets must be in orbit around Jupiter, under bombardment by particles from a source yet to be identified. That, in turn, is responsible for the permanent renewal of the rings. Not until mid-1996 did the "long-sought breakthrough" occur. For the first time, physicists were able to include all aspects of the ring system's environment in their calculations. The critical element involved the behavior of electrons in

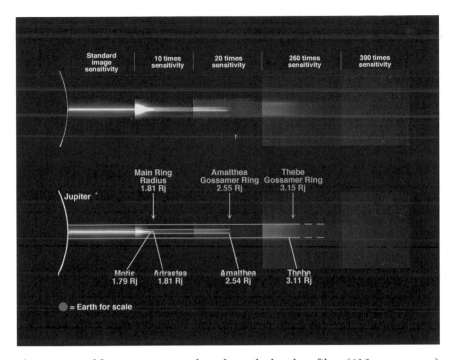

This mosaic of five images was taken through the clear filter (610 nanometers) of Galileo's camera on October 5, 1996. The mosaic is shown twice; the top panel displays only the data, while the bottom panel gives the location of some of Jupiter's small ring moons and presents a match between the image and a simple geometrical model of the gossamer rings. From the spacecraft's distance of approximately 6.6 million kilometers, the images have a resolution of about 134 kilometers per pixel. They were acquired when Galileo was in Jupiter's shadow, peering back toward the Sun, hidden behind the planet. The spacecraft was located only about 0.15° above the ring plane at the time, making the images highly foreshortened in the vertical direction. North is to the bottom.

Jupiter's ionosphere. It turned out that estimates of the life span of the individual particles making up the rings shrank from approximately 100 years to less than one year. And the smaller particles disappear faster than the larger ones, sometimes within a matter of days! With this information researchers were able to calculate the size distribution of the matter making up Jupiter's ring system and model its appearance.

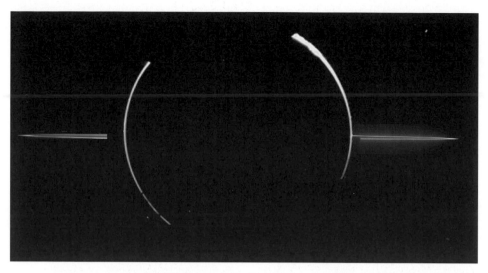

Jupiter's rings in perspective: This combination of several individual pictures by the Voyager and Galileo spacecraft shows what Jupiter would look like to a keen-eyed observer with the Sun behind it. Only then do the narrow and faint rings become obvious.

All that scientists were missing now was the source of the particles colliding with the moonlets, and here as well they found the answer. The size distribution of the incoming particles had been determined— the big ones were vastly outnumbered by the small ones. If it was not interplanetary dust that made up the rings, it could only be ash from Io's volcanos. Finally, 17 years after the Voyagers first delivered the data, a provisional conclusion had been reached—and once again fascinating Io was part of the process.

In the mid-1980s, image analysts discovered in the Voyager pictures another extremely weak extension of Jupiter's rings, this one out into space. Called the "gossamer ring," it reaches out at least as far as the orbit of the small moon Amalthea. The role played by this ring was not immediately clear, but it corresponds to a similarly delicate extension of Saturn's main ring. Understanding Jupiter's ring system in detail

The best picture of Jupiter's gossamer ring, backlit from Galileo's perspective in Jupiter's shadow. The orbiter was located practically inside Jupiter's equatorial plane, causing the ring to appear as a single band, but also making it more visible. Jupiter itself must be imagined as being located about as far to the right as the size of the image.

became another of the tasks assigned to Galileo. With its CCD camera, it would be capable of much higher resolution than the Voyagers.

The orbiter shot its best pictures of Jupiter's rings in November 1966. From just 0.5° above the plane formed by rings, the shots were taken from outside Jupiter's shadow, so that the rings were backlit. Two new aspects of the rings became immediately apparent. For the first time, it became possible to discern a pronounced radial structure in Jupiter's main segment, almost like Saturn's ring. The thickness varies considerably, moving from the outside in toward the planet. Probably responsible for this effect are small (invisible) moons. Although the ring is made primarily of small particles very much under the influence of Jupiter's magnetosphere, the definition of the extreme outer edge of the ring is remarkably sharp. New insights came especially from Galileo's long exposures, which clearly showed all three components of the ring system—the flat main segment, its lens-shaped halo, and the gossamer ring, the halo's diffuse extension out into space. The torus-shaped halo is unique in the Solar System, probably the product of electrostatic forces that lift the tiniest particles as much as 27,000 kilometers above the plane of the ring. The little-understood gossamer ring is clearly visible in these photographs for the first time. It appears to be a disk of constant thickness and density, extending far into space beyond the main ring. The modelling of Jupiter's ring system made yet another advance in late 1998 when new Galileo images with high sensitivity became available. Looked at exactly from the ring plane the rings actually look like boxes—and their respective edges correspond to the orbits of the four innermost small moons of Jupiter. Only the new Galileo images revealed, for example, that the gossamer ring was actually two nested rings. The outer, fainter gossamer ring reaches to a distance of 221,000 kilometers from Jupiter's center, right to the orbit of Thebe. And the brighter, inner one extends to 181,000 kilometers or Amalthea's orbit. This practically proves that these two moons are the main sources of the rings. And the thickness of the rings turns out to be a simple consequence of the inclination of the moons' orbits. As suspected for a while, the much brighter main ring is linked to the other two known inner Jovian moons, Metis and Adrastea. No new insights into the origin of the torus-shaped dust halo around the ring system were gleaned—the leading model remains the electrostatic charging of ring particles by the magnetosphere which then drags them away. The same processes that have now become surprisingly "obvious" in the Jupiter system should also work at the other gas planets with their many moons.

And even that is not the end of the story! Computer models showed in 1998 that Jupiter's magnetosphere should be capable of cap-

turing dust particles that enter the Jovian system from the outside, be it
from interplanetary or even interstellar space. These particles have traveled in the interplanetary medium for a long time, and the solar wind—
charged particles streaming at high speed from the Sun—should have
given them a positive charge. Not neutral anymore, they feel the mighty
Jovian magnetosphere. The numerical simulations showed that Jupiter
should be able to catch a considerable number of these particles—and
force the majority of them into retrograde orbits, i.e., they should orbit
Jupiter in the opposite direction as all the moons and the particles in
the known rings do. Has Galileo actually detected this hypothetical
new Jovian ring? There is no direct proof, but the spacecraft's dust

A schematic view of Jupiter's various rings and their relationship to the orbits of some
of the small moons.

detector has repeatedly seen individual particles in retrograde orbits whenever it approached Jupiter to less than 20 planet radii. And we will never be able to image the weird new ring—according to the models there will be one single particle, sized between 0.05 and 1.5 meters, in a cube 450 meters on one side . . .

Jupiter's Small Moons

The Galilean moons are only the most obvious part of a constellation of no fewer than 16 Jovian moons that have been discovered so far. Four small moons circle the planet inside Io's orbit, and another four circle in the same direction outside Callisto's orbit. The four with the largest orbits are going *backward* around Jupiter. All eight outer moons are probably asteroids that used to revolve around the Sun until they were caught on a near flyby and sidetracked into a Jupiter orbit. The only reason the four most distant moons, Sinope, Pasiphae, Carme, and Ananke, remain in their orbits is because they approached the planet from the "wrong way around." Gravitational disturbances caused by the Sun would otherwise have driven them out of the Jovian system a

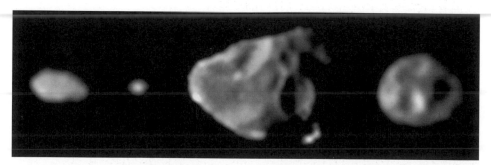

Family portrait of Jupiter's small inner moons. In approximate scale to each other, left to right, are Metis (maximum span about 60 kilometers), Adrastea (20 kilometers), Amalthea (274 kilometers), and Thebe (116 kilometers). The Voyagers had been able to see details on the surface of Amalthea, but the other moons were known only as points of light prior to Galileo's arrival.

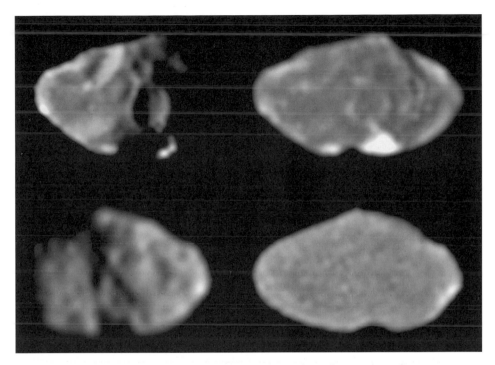

Four views of the Jovian moon Amalthea, with north at the top in each image. Twice the Sun was shining from the side, and twice from the front, accentuating albedo variations, i.e., actual differences in the brightness of the surface, as opposed to the effects of light and shadows.

long time ago. As it is, it takes them about two Earth years to accomplish their 20- to 24-million-kilometer revolution around Jupiter.

It is also conceivable that Jupiter captured only two moons initially, and they collided, breaking apart and forming two pairs. Before the Voyager mission, the only moon known inside Io's orbit was Amalthea, a dark red, very irregularly shaped body, about 270 kilometers long and 150 kilometers in diameter, with its long axis pointed toward the planet. Several impact craters are visible, including an especially large one called Gaea on the moon's south pole which is filled with some kind of shiny material. Between the orbits of Io and

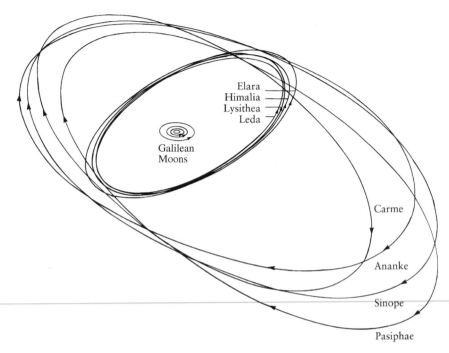

The outer moons travel in very eccentric orbits around Jupiter, and their orbits are also tilted in relation to Jupiter's equatorial plane. The four inner
moons in this group circle Jupiter in prograde orbits at distances between 11
and 12 million kilometers. The outer four, at distances of 20 to 24 million
kilometers, move in a retrograde (backward) direction around the planet.
There are no close-ups of any of these moons.

Amalthea, the Voyagers discovered Thebe, about 80 kilometers across,
and, in the ring's outer regions, the much smaller Adrastea and Metis
(40 and 25 kilometers, respectively). They race around Jupiter in a
third of an Earth day, at a pace faster than the planet rotates. An observer who took up a position on Jupiter's clouds, were that possible,
would see Adrastea rise in the west and set in the east. It is known
for certain that the Galilean moons and the four innermost moons
were formed at the same time as Jupiter, constituting along with the

gas giant a miniature planetary system (see pages 252–261 for a full discussion).

A New View of Jupiter's Dust Streams

No fewer than three completely different "populations" of dust particles can be found in, and sometimes leaving, the Jovian system. That is the conclusion drawn by scientists from data delivered by Galileo's dust detector, which was made in Heidelberg, Germany. Extraordinary findings were already coming in from Galileo's initial approach to the planet, and on the orbiter the observations naturally continued. The detector is sensitive to particles from about 1/100 to 10 microns in size and, in contrast to nearly all other Galileo instruments, has been turned on almost constantly since the launch in 1989. Still, the real suspense came only after Galileo's arrival, starting with an enlightening Io flyby on December 7, 1995. For the whole first week of December, about 150 extremely small particles a day were hitting the detector, just barely enough to be taken seriously. As soon as Galileo passed by Io, these detections ceased—precisely what scientists expected if the source of the particles was Io. The dust detector's "viewfinder," while generous in size, was not unlimited, and after the flyby Io had moved out of its range of vision.

The impact rate of a much rarer kind of micron-size particle steadily increased, however, as Galileo neared Jupiter. These particles obviously were in orbit around Jupiter, and could be of local origin. In that case the source might be the outermost small moons, the Galilean moons, or Jupiter's ring. It was also possible that they had come from interplanetary or even interstellar space and had become concentrated in this area by Jupiter's powerful gravitational force. Subsequent orbits around Jupiter were supposed to settle some of the questions, but unfortunately the dust detector's operational time was limited. In any case, the impact rate of particles measuring less than a micron varied considerably, and it varied with suspect regularity, every five and ten hours, corresponding to Jupiter's half- and full-rotation periods. And the impact rate varied precisely with Galileo's position over the equator of the magnetic field.

Obviously, Jupiter's magnetic field was exercising a heavy influence on the orbits of this dust population—it was exactly the same dust that Ulysses and Galileo had begun encountering a great distance from the planet. By now there was no doubt whatever that the particles in this "dust stream" carried an electric charge and had been hurled out of the system by Jupiter's magnetic field. A precise analysis of the particles' angle of impact located their source at a distance of five to eight Jupiter diameters away from the planet. But Galileo could not identify any dependence between impact readings for this kind of dust and the positions of the Galilean moons. It was not, in any case, directly from the surface of the moons that the particles got started on their spiral out of the system. Instead, the remarkable journey originates from a more *extended* source—possibly in the Io torus or the gossamer ring.

Otherwise Galileo registered a concentration of small dust particles in the immediate vicinity of Ganymede. On both flybys, only a couple of particles hit the detector, but each time the impacts came within minutes just before and after the closest approach. The particles could be microscopic fragments split off from the surface of the moon by other particle collisions. And the detection rate briefly peaked on the first Callisto and Europa flybys. Clearly, all the Galilean moons are sources of split-off "secondary dust." The third dust population, particles in the micron range which Galileo had already encountered on its visit to Io, were likewise there on the Ganymede, Callisto, and Europa flybys. They appeared indeed to be revolving around the planet in orbits of less than ten Jupiter diameters—and their exact origin remains a mystery.

Progress Report: Io Makes the Dust Streams

A breakthrough in understanding the origin of Jupiter's dust streams finally came in 2000—by a mathematical analysis of the impact rates in Galileo's dust detector as a function of time. A frequency analysis of data from 1996 and 1997 revealed several strong periodicities. There was Jupiter's rotation—and there was the orbital period of Io! This moon, or rather its volcanos, was thus the source of the tiny dust particles which would later be ionized and forced by Jupiter's powerful magnetic field to follow the planet's rotation. Other sources of the dust that had still been a possibility, namely Jupiter's main ring, the gossamer ring or dust left over from comet Shoemaker-Levy 9, can now be ruled out. Actually the implication of Io as the dust source was no surprise. About one ton of material per second escapes through its volcanic plumes. The fact that it is the volcanos and not micrometeorite impacts onto Io's surface that produce the dust is clear because of the small size of the dust. The particles making up the dust streams are of submicrometer size, and the size of particles that can escape from a typical Io volcanic plume is 0.01 microns or less. Other studies have also shown that impact ejecta are way too inefficient to be a dominant source for the dust streams.

"Dust from Io's volcanos is a minor dust source, compared with collisions of the main belt asteroids or comet activity," write the Heidelberg dust researchers. "Nevertheless, it adds to the variety of dust sources in the Solar System. At a velocity of 200 kilometers per second or more, the Jovian dust stream particles can also leave the Solar System to slightly populate the local intestellar medium. We can now use dust stream measurements to monitor Io's volcanos' plume activity. Such measurements are a unique complement to the partial glimpses acquired by Galileo and ground-based image observations, because our temporal coverage is more complete (we provide information for each Galileo orbit), and because the DDS measurements give an estimate of the integrated total amount of volcanic material from Io's more than 100 volcanos, for example—material that escapes from Io, and which then disperses through the Jovian system painting the other satellite surfaces. This discovery also lends support for using dust measurements as a probe for charging effects in the Jovian magnetosphere. Dust from the dust streams is clearly magnetically controlled. Dust particles carry information about charging processes in regions of the Jovian magnetosphere, where information is otherwise sparse or unknown."

Callisto, the Outsider

Callisto is the outermost of the four Galilean moons. It is the only one for which the reputation shared by all of them in the 1970s—that they were primitive and inactive bodies—has been borne out. We have seen that this assessment does not hold for the other three, whereas Callisto does indeed appear to be a relic of the distant past, just as expected. The surface of the 4,840-kilometer moon is extremely old and scarred by countless impact craters, because there is no geological process under way to erase the signs of past collisions. Callisto is in fact absolutely unique among all planetary bodies covered with impact craters, because it has no "plain" regions, where the craters have been covered over by more recent processes. Nor does Callisto generate a magnetic field or show any sign of having its own magnetosphere. In terms of internal composition, readings of Callisto's gravitational field show it to have none of the usual differentiation between a dense metallic core and a crust. It is made of a homogenous blend of 40 percent ice and 60 percent rocky material, as well as iron and iron sulfide.

Callisto nevertheless has its charms. It makes for a very good study of long chains of impact craters, which also makes it important for comet research. The assumption is that the moon has undergone catastrophes in the past not unlike what happened to Jupiter itself from 1992 to 1994. That was when tidal forces near Jupiter broke the comet Shoemaker-Levy 9 into a number of fragments, stretching out like a string of pearls in space before crashing one by one into the planet two years later. It would appear from the lasting chains of comet ruins on its surface that Callisto has been a frequent witness to the breaking up of comets like this.

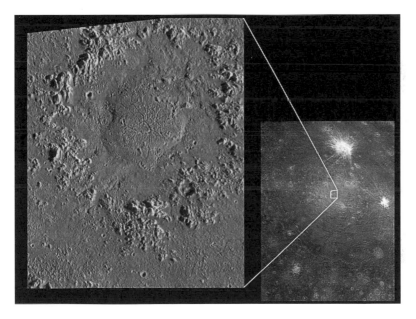

Here data from two orbits showing several types of impact craters are combined. The global image on the right shows one of the largest impact structures on Callisto, the Asgard multi-ring structure with a diameter of 1,700 kilometers. It consists of a bright central zone surrounded by discontinuous rings. Smaller impacts have smashed into Callisto after the formation of Asgard. The young bright-rayed crater Burr located on the northern part of Asgard is about 75 kilometers across. A third type of impact crater is represented by Doh, located in the bright central planes of Asgard and 55 kilometers in diameter. Doh is remarkable for its 25 kilometer big dome in the center where one normally finds a bowl-shaped depression or a central peak. The presence of the dome could mean that Asgard's formation has turned the underground into slush so that the mechanics of subsequent impacts change.

It takes a detailed study of Callisto's craters to note the differences between them and the heavily cratered areas on our own Moon or Mercury. Callisto is essentially a ball of ice, with a mean density of 1.8 grams per cubic centimeter, rather than a body made primarily of rock. Temperatures never rise above $-130°$ Celsius anywhere on the surface, and underground they are more in the area of $-170°$ Celsius. The way water ice behaves at such temperatures is practically indistinguishable from the behavior of rock. Over geological time spans, however, water ice has a slight flow, like a very slow glacier, causing the shape of the craters to change noticeably. Craters of all sizes on Callisto are much flatter than their counterparts on rocky planets. The biggest impact basins are neither deeper in the center nor circled by mountain ranges. Instead of structures of this sort, which are typical for the Moon or Mercury, Callisto has a number of large bright cross-shaped formations. These structures are comparable in size to more conventional basins, but they have scarcely any topographical relief. Over millions of years the contours have flowed away, which is one of the most important indications that Callisto's crust is made of water ice.

Callisto has one final major distinguishing feature, having to do with a soft, dark material that covers the icy mountains and small craters—which, whatever it is, is not ice. It could be that rocky material is blended through Callisto's ice. And if the ice components are very gradually disappearing into space, whether because of sublimation (the transformation of individual molecules from a solid state directly into a gas) or from being bombarded by particles from the magnetosphere, then what is left behind becomes increasingly rocky. Another puzzle is why observers find hardly any impact craters less than 10 kilometers in diameter. Either Callisto's dark coating is younger than most of the craters—which requires a mysterious process that could have produced it so quickly—or it is constantly in motion, so that small craters are quickly destroyed. Of the four major moons, Callisto is the farthest

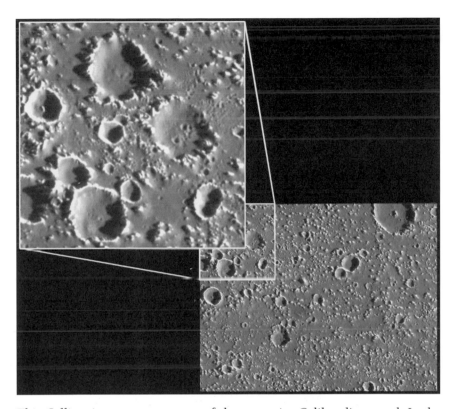

This Callisto image presents one of the mysteries Galileo discovered. In the upper left corner of the bottom image, what appear to be very small craters are visible on the floors of some larger craters as well as in the area immediately adjacent to the larger craters (see enlargement). Some of these smaller craters are not entirely circular—and perhaps they are not impact-generated after all but the expression of some endogenic process, i.e., some surface or subsurface phenomenon. Then again they could be just eroded secondary craters, formed when an initial large impact ejects large enough pieces of the surface that the pieces themselves create small craters.

away from Jupiter, and it may be that it lacks any notable interior turbulence. Even for Callisto, however, the possibility of certain kinds of activity on a smaller scale cannot be ruled out. And there are indications of a thin atmosphere. An analysis of plasma waves in its surroundings showed it to be "a significant source of locally produced

Galileo and Solar Research

It had been known for decades by solar researchers that there are two types of solar wind, both of which can be measured directly by probes in interplanetary space. We distinguish between the "slow" component, which travels at a speed of less than 500 kilometers per second and is very diverse in its physical characteristics, and the fast solar wind, moving at an average of 750 kilometers per second and quite constant in its physical parameters. Lacking was a clear identification of the sources of the two types of wind. Astronomers assumed that the slow wind came from the spectacular Helmet Streamers (named for their distant resemblance to a medieval spiked helmet), which are the most conspicuous of all solar formations to the eye. The Sun's magnetic field is responsible for their characteristic narrowing toward the outer edges. Astronomers believed that the slow solar wind flowed precisely out of this furthest extension of the Helmet Streamers, the "streamer stem." And they expected to find the source of the fast solar wind just in those places where the corona is least conspicuous, or even where it displays some extensive gaps. These "coronal holes" are especially common over the Sun's polar regions, but can at times extend as far as 40° toward the solar equator.

New data in early 1997 from Ulysses, a probe launched by the European Space Agency, caused scientists to extend the source of the fast solar wind to the entire surface of the Sun, except of course for the places where streamers are located, which produce the slow wind. Readings from various parts of the corona were unambiguous. The speed of the solar wind is higher between the streamers than inside them, and the boundary between them is sharply drawn. This is where Galileo entered the picture. In January 1997, with Jupiter almost exactly behind the Sun from the perspective of Earth, came the opportunity for an unusual experiment. Galileo's radio signal would pass through the corona, the extended hot solar atmosphere. This bonus gained from Galileo's radio science completed the picture; in readings taken from January 15 to February 4, 1997, a "scintillation"—a tiny change in the "pitch" of the transmitter's radio signal—was confirmed as the waves passed through the plasma in the corona.

Most of the time, the scintillation was minor, but on January 22 in particular there was a major variation—just as the radio signal went through the middle of a prominent streamer stem. This finding confirmed expectations that the slow solar wind really does originate in streamer stems. That the scintillation in Galileo's signal over the three-week period was otherwise nearly insignificant confirms once again that the entire corona is dominated by the fast solar wind, and that its source is not at all restricted to the coronal holes. So theorists working on questions about the Sun likewise had reason to be grateful for the multifaceted Galileo.

1997/01/17 22:18:32

The solar corona on January 17, 1997, taken by the SOHO (Solar and Helio-spheric Observatory) coronagraph LASCO C2. The searing hot disk of the Sun and the areas immediately around it are masked by a shutter (indicated by the circles), but the more extended parts of the solar atmosphere, espe-cially the Helmet Streamers, are clearly visible. The bright spot to the lower left is Jupiter, at this resolution practically identical with the location of Galileo. Indicated by the black lines are the locations of the corona where SOHO's UVCS (Ultraviolet Coronagraph Spectrometer) took readings of the velocity of the solar wind. In February 1998 this conjunction of Galileo, the Sun, and Earth was repeated—and almost simultaneously there was a total solar eclipse, offering astronomers even more possibilities for observing the Sun's corona.

plasma." The more data from Callisto came in, the more interesting the moon became. NIMS spectra confirmed the existence of an—albeit extremely thin—atmosphere made up from carbon dioxide and possibly other species. All four Galilean moons are now known to have tenuous atmospheres, being replenished continuously by particles leaving their surfaces. The biggest surprise, however, were clear indications of a sub-surface ocean on Callisto, just as it is hypothesized for Europa! Callisto affects the Jovian magnetic field in a similar way, and so the existence of a salty ocean has been postulated for Callisto as well. Continuing analyses of the gravimetric data also indicate that this moon is not the completely undifferentiated body that everyone believed it to be. While Callisto certainly has not separated into a metallic core, a rocky mantle and an ice-rich outer shell as the other three big moons, it is kind of "half-baked." Since Callisto never in its history has experienced the tidal heating effects that have transformed the other moons (and are still working on Io today), its ingredients are somewhat separated but still largely mixed together. So why might Callisto have a liquid inside even today? Radioactive heat might be an explanation.

Jupiter and the Four Galilean Moons—A Planetary System in Miniature

Having looked at each of the four major moons independently, it is time to shed a bit more light on the system as a whole. Each of the Galilean moons has its individual nature, with no two bearing even a passing resemblance to each other, and yet the four moons make up the most clearly organized planetary system we know. The two innermost bodies, Io and Europa, are about the size of our own Moon and are both made primarily out of rock and metal. In addition, Europa is equipped with a remarkable outer covering of ice. Ganymede and

Mission Accomplished: Galileo's Technological Balance Sheet

On December 7, 1997, exactly two years after first entering orbit around Jupiter, Galileo's primary mission ended. Research would by all means continue, but there was reason enough for Galileo project managers and scientists to seek an accounting, to attempt to make their accomplishments apparent. Things had looked very different indeed five years earlier, when scarcely anyone was prepared to believe the assurances of JPL scientists that 70 percent of mission objectives could be fulfilled even without the high-gain antenna (see page 96). The only hope had been that scientists would be able to choose, among countless possibilities, precisely those observations that would be the most informative—true to the dictum that not all data bits are alike and that there was no reason why the sheer number of ones and zeroes transmitted necessarily determined the amount of valuable *scientific* information that could be gained. Anyone who kept up with the succession of images on the Internet, the numerous press conferences, or the rising flood of specialized publications could easily forget that it was all coming from a practically mute spacecraft.

"The Galileo orbiter is still working flawlessly, and all scientific instruments continue to deliver excellent data," announced outgoing project manager Bill O'Neil on November, 5, 1997, in a statement. "The scientific yield is tremendous, and the analysis will certainly add many important discoveries. Even though the antenna problem has caused us to miss some of the original goals, I am sure that the scientific results easily exceed what we had hoped to accomplish when the project began 20 years ago." In the usual procedure whenever a project milestone has been reached, NASA named a new project manager for the "Galileo Europa Mission." O'Neil, who had shared all the ups and downs of the project since 1977, was replaced by Bob Mitchell in December 1997.

Mitchell can look back on a career with the Galileo project almost as long as O'Neil's. From 1979 to 1988, as Mission Design Manager, he led the Galileo mission through the multiple revamps of the 1980s. In particular, Mitchell oversaw the working group responsible for developing the complicated VEEGA route, without which Galileo would never have made it to its destination (see page 44). In an interview on December 15, 1997, Mitchell took stock of the primary mission and the state of the orbiter's health as it began the extended mission to Europe. "Generally very good" was the judgment in regard to the entire array of scientific instruments, eight years now since the launch. One essential sensor in the Plasma Wave System had failed, but only upon its ninth encounter with a moon, when practically all the data had already been recorded—basically a marginal impact on the mission. Other instrumentation problems had even been solved—a filter that became stuck on the Photopolarimeter, for example. Once it was freed, scientists simply made sure not to move it back through the critical position, and ever since it had followed commands flawlessly.

There was no reason to complain about the tape recorder in 1996 and 1997 either. Through a series of tests, technicians had characterized its behavior and weaknesses—circumstances in which the tape sometimes got stuck—so well that it had never failed again. And so there was hope, after the completion of the extended mission to Europa and Io at the end of 1999, that the recorder could be rewound all the way to the start, allowing scientists to play back Galileo's first photograph of Jupiter upon arriving in 1995. Since then, there had been no chance for Galileo, with its single video camera, to take another picture of the planet as a whole.

The balance for Galileo's electronic data storage system was also positive. By the end of 1997, models had predicted memory cells failing by the dozen, but as it turned out failures were limited to exactly one cell. The vulnerability of Galileo and its systems to radiation was surprisingly slight, with the gradual degradation of many instruments still to come.

For flight controllers, too, Galileo had been remarkably good natured its entire time in Jupiter orbit. The onboard computer had sometimes gotten "stuck" when new software was loaded before having been perfected, but that comprehensive state of emergency in which Galileo could be "frozen" for days at a time had not happened a single time. The small German-made thrusters, used for attitude control and for changing the spacecraft's trajectory, were ultimately a success as well. In 1988, the decision had been made to operate them exclusively in pulse mode to prevent overheating. They were never fired continually, as they had actually been designed to do. That caused fuel consumption to rise *relative* to original specifications—but at the same time, the thrusters' efficiency rating was clearly toward the top of predicted performance standards. And that, combined with fuel supply estimates that had always been pessimistic, led in turn to an increasingly positive fuel consumption picture, allowing for two asteroid detours, all of the encounters planned for the primary mission, and extension after extension of the project. It can be said with no exaggeration that Galileo remains fit for further adventures in space.

Callisto are bigger, on approximately the scale of the planet Mercury, and almost half of each one (by mass) is made up of ice. This distribution recalls that of the Solar System as a whole, which likewise has inner planets with cores made of rock and metal and much larger planets composed largely of gases and ice farther out.

Jupiter's moon system is even more systematic, however, in that most critical characteristics vary continuously with distance from the planet. All four moons may be solid bodies, but the share of ice in overall composition increases steadily from the inside out. Io has no water ice whatsoever, whereas Europa has at least an ice crust. On Ganymede and then truly on Callisto, water ice is the dominant material. Geological activity follows the same pattern as chemical composition. Io is all but hyperactive. Europa may be active at present and certainly was in the past. Ganymede went through turbulence at least at one time in the distant past. Callisto seems not to change.

For the same reasons, Callisto is marked by innumerable craters, and cratered regions can also be found on Ganymede. Europa has an individual crater here and there. And Io has none at all, even though the gravitational force of gigantic Jupiter works effectively to send passing comets and asteroids Io's way. Collision rates must be much higher here than they are on Callisto.

Of course Jupiter itself is responsible for all of these diverse characteristics. The gas giant and its four largest moons came into being together, taking form out of the early solar nebula, the disk-shaped structure with the Sun forming in the center and the other planets around it. As the proto-Jovian body was taking shape, this large-scale process was repeated in miniature. Here as well, an excess of angular momentum created another disk, with Jupiter at the center and the moons forming around it in a planetary system of their own. The parallels between Jupiter and its moons and the Sun and its planets was even more striking in the beginning. According to computer models, Jupiter started off both bigger and more *luminous* than it is today, which means that the inner regions of the proto-Jovian cloud remained fairly hot. Only rock and metal could condense out of the cloud here, and no water ice—which is why Io has none at all and Europa has only a crust. As Jupiter developed, it went on exercising considerable influence over the moons, kneading them with its tidal force and heating them up.

The tides are so powerful that they can be responsible for more local heat production then radioactivity from the uranium, potassium-40, and thorium present in the rocky parts of the moons. They have this effect, however, only because the moons do not orbit in perfect circles. The tendency of the tidal forces is to pull the orbits into the shape of a cross, while the gravity of the other moons works to counteract that tendency. This is caused by the "resonance" that exists among the orbital periods of the moons. It takes Europa 3.6 days to go around Jupiter, double the 1.8 days it takes Io, and Ganymede's 7.2-day orbit is once again double Europa's. In this way, the positions of the moons in relation to each other are reproduced regularly, which stabilizes the whole system—including the way the individual orbits deviate from a cross shape. This phenomenon is called a "Laplace resonance," and it is unique in the Solar System. In all likelihood, the Galilean moons drifted into this special configuration over the course of time.

This change in the moons' orbits relative to each other, which may have taken place even billions of years ago, can be explained in terms of celestial mechanics. Jupiter's tidal effects also work the other way around, of course, with the moons causing minimal "high tides" on the gas giant, which in turn transfer angular momentum back to the moons. And that, finally, has the effect of slowly increasing their distance from Jupiter. This same process continues today in the Earth–Moon system. The difference is that eventually, Io, Europa, and Ganymede settled into a Laplace resonance. Io moved away from Jupiter faster than Europa, and Europa faster than Ganymede. At some point in the development the numbers were just right, and the moons were captured. Only Callisto maintains its distance from all these resonance affects, and from this perspective it is the outsider in the group.

Tidal effects increase with nearness to Jupiter, driving Io's ceaseless volcanism, whereas on Europa they suffice only to maintain its extraordinarily flat and bright ice surface. Ganymede, as the heaviest and most distant member of the resonance system, operates in a certain sense as an anchor. Subject *at present* to no tidal effects worth mentioning, that need not always have been true.

Computer simulations of the orbits of Jupiter's moons show that the eccentricity of Ganymede's orbit could have been considerable at one time. Tidal forces also would have been more powerful then, suggesting that this would have been when the tectonic and volcanic history of the moon, evident on its surface, took place (see pages 148–151). Only Callisto, with its near-perfect circular orbit beyond the reach of Laplace resonances, never underwent any extra warming of note.

None of these considerations are new, having begun with the Voyager missions, and they can account adequately for the way the individual Galilean moons look so different from the outside. Nevertheless, the final questions remained for Galileo to answer. As part of its primary mission the orbiter ventured close enough to each of the four major moons to use its magnetometer and plasma detector to study their internal magnetic fields. Since Galileo came so much closer than the Voyagers, the effect of the moons on Galileo's trajectory was large enough for scientists to follow by precisely monitoring changes in the orbiter's radio signal. The moons' internal structures became accessible for the first time and were no longer left to mathematical models based inevitably on assumptions.

Any circularly symmetrical distribution of mass will have the same center of gravity as a point of mass, as Isaac Newton had already calculated in the seventeenth century. When the symmetry is disturbed, however, by rotation or tidal forces, the gravitational field changes in predictable ways. Even the flight of a spacecraft through such a no longer symmetrical gravitational field is enough in principle to allow scientists to calculate the heavenly body's inertial moment. There remains yet a long way to go, to be sure, from there to a complete understanding of its inner structure. However, a few assumptions about the physical and chemical behavior of the metallic, rock, and ice components are permitted. They are quite possibly structured in concentric shells, and density will decline steadily from the inside out.

The findings from Io, Ganymede, and Europa were unambiguous: all three moons are centrally condensed. In exactly the same way as Earth-like planets, they have separated or differentiated internally into various layers. Io has a metallic core and a rocky mantle, which is no great surprise given the intense tidal heating to which it is subjected. Ganymede is different. Differentiated with exceptional precision, it has an 800- to 900-kilometer ice mantle covering a core made of iron and rock. And there is no layer in between in which the ice and iron would be mixed together. Moreover, Ganymede has a rather strong dipolar magnetic field. Although Europa is also centrally condensed, with an outer surface of ice 100 to 200 kilometers thick, a rocky layer, and probably a metallic inner core, some of the data regarding this moon were contradictory. Proximity to Jupiter hampers the search for a magnetic field, but it seems that an internally generated dipolar field is possible. And with Callisto, the view is currently changing from an ancient inert body to a surprisingly evolved—and perhaps even somewhat active—body.

It is not just the physics of the varying densities of materials that determines whether celestial bodies differentiate. Chemistry also plays a role. It sounds very simple: heat up a rocky core and the metal will melt and collect in the center. But the rocky core of a moon made substantially of ice forms when the ice melts and the rocky components sink to the center, and a core of that sort is necessarily highly oxidized. Under pressure and with increasing temperatures in the core, and also because of the presence of carbon, all iron oxides react with iron-

The Scientific Balance Sheet: The Biggest Discoveries

Not only did the Galileo spacecraft turn in a flawless performance technologically (aside from the high-gain antenna), but the mission can also be called a great success because of the scientific discoveries that were made.

- The most important finding of the entire two years probably concerned indications that there might still be *liquid water under the ice crust of Europa.* Images with a spatial resolution of 50 meters show icebergs or "ice floes" in a frozen sea. The ways in which these individual floes have broken apart and shifted in relation to each other practically constitutes proof in its own right of the previous activity of sea currents, for Europa has neither downward slopes on its surface nor winds that could have caused the changes. The small number of impact craters further indicates that this part of the surface is very young, maybe as young as 25 or 30 million years. Considering that the moon as a whole is 4.6 billion years old, because it was formed at the same time as Jupiter, it is basically inconceivable that all of the water could have been completely frozen from the start.

- The orbiter and the atmospheric probe explored Jupiter's atmosphere *as a team,* with the results gathered from a distance providing the perfect complement to the readings made firsthand by the probe. Only because of the NIMS data do we know, for example, that the amount of water vapor in the atmosphere varies by orders of magnitude. Had only the probe been sent to Jupiter, none of the findings—about the clouds and the chemical composition of the entire planet—would be representative. There is no doubt that the most important observations on a small scale had to do with the existence of clearly defined storm cells.

- Most interesting in regard to Io were the changes in the moon's volcanic activity since the time of the 1979 Voyager encounters. Monitoring Io at some point during every orbit, Galileo identified at least 19 active volcanic centers. Both the presence of silicate volcanism and occasional very high lava temperatures were confirmed, suggesting extreme types of chemical reactions. In total, however, Galileo saw fewer eruption clouds than the Voyagers in 1979. Are they still there, but with their visible particles depleted? On six different occasions, Galileo transmitted its radio signal through Io's complicated ionosphere, exploring it in unprecedented detail. Measurements of the moon's gravitational field on the first day of Galileo's arrival imply a large iron or iron sulfide core, occupying about half of Io's radius.

- Europa also has an ionosphere, along with a three-part internal structure: a large metallic core, a rocky mantle, and an outer shell of ice and liquid water, perhaps 150 kilometers thick. Ganymede, too, is internally differentiated, and it even has its own magnetosphere—the first ever to be discovered on a moon. On Ganymede, Galileo also discovered a thin hydrogen atmosphere and a powerful outflow of protons. Homogenous and with no fields of any kind, Callisto was interesting ultimately for the continuing puzzle posed by its cratering history.
- Galileo's nonimaging instruments studied the large plasma torus in Io's orbit, the geometry of which could change sometimes in the course of a single day. The variations may be caused both by differences in how much it is being charged by material from Io and by changes in the solar wind. Jupiter's polar lights are much brighter at some times than at others, due to the attraction of streams of particles into Jupiter's atmosphere. Galileo's particle detector was actually able to gauge such streams.
- On the whole, the patterns of radiation in Jupiter's magnetosphere have undergone dramatic changes since 1979. The density of the hot plasma has decreased, and the composition of the ions is different today, caused perhaps by changes in volcanic activity on Io. The answers to this and many other questions raised by the Galileo mission are presumably to be found in the masses of data that—despite the failure of the high-gain antenna—have piled up in excess of original projections. No visits at all to either Io or Europa were foreseen in the original plans, for example.

The success of the Galileo mission amounts to one of most astounding improvisations in the history of space exploration, comparable perhaps only to the extraordinary story of the Hubble Space Telescope, as it went from scientific disaster to world-wide fame.

containing silicates, leaving behind iron sulfide. Given the pressure at the core, the melting point of iron sulfide is over 1,200° Celsius. And that means that differentiation simply cannot take place. Before the core gets this hot, the rocky material starts wandering back out toward the surface, counteracting any further rise in temperature—with the result that metals and silicates do not separate.

This is only one of many mysteries the Jovian system presented to its explorers while the Galileo mission continued. Whenever better data became available about some detail, cherished models became endangered. One prediction, however, did come true. By undertaking a systematic study of the Jovian system—as only the Galileo mission made possible—we have become a great deal more aware of what we know about the history of the Solar System as a whole. And what remains to be learned.

Chapter Five

What the Future Holds in Store

Galileo Flies On

For two years the Galileo spacecraft had been weaving a path through the Jovian system, accomplishing 11 flybys of Jupiter's moons. The mission was scheduled to end in December 1997, but already in the summer of 1996, even before Galileo had really gotten into gear, discussions had started about prolonging operations. In early 1997, the extended mission won approval. There was, however, a down side. A two-year extension had been possible only on the condition that operating expenses be drastically reduced. NASA had only $30 million total for the Galileo Europa Mission (GEM), scheduled to last from December 7, 1997, to December 31, 1999. The team that had been overseeing the Galileo project thus far was cut by 80 percent, and everything for which it was remotely possible had to be simplified and automated. Now, for example, data could be gathered on moon flybys for only two days, rather than the previous seven, and the routine monitoring of Jupiter's magnetic environment was stopped almost completely. The reduction in staff to 20 percent of previous levels also meant that the experts needed to deal with unforeseen developments would find work on other projects. Only in case of a real emergency would they be called back as "tiger team" members.

As already implied in the name, the main goal of the Galileo Europa Mission was the further systematic exploration of the moon Europa, although Jupiter was kept in mind, as well as Io at the very end. The camera was to take about 1,000 additional photographs in the infrared and ultraviolet spectra, and new field and particle data were expected. The expanded mission had three phases. The first included eight orbits around Jupiter, with close Europa flybys, lasting from December 16, 1997, to February 1, 1999. Then Galileo orbited Jupiter four times, during which its orbit was steadily lowered (reducing the perijovium), making it possible simultaneously to explore Io's plasma torus and details in Jupiter's clouds. Finally, two more close Io flybys were planned between October 11 and November 26, 1999, assuming that Galileo's onboard electronics could withstand the extreme radiation it encountered in this part of Jupiter's magnetosphere.

The eight Europa flyovers (with altitudes ranging from 3,600 to a mere 200 kilometers) had primarily to do with continuing the search for possible ongoing geological activity and for clearer indications of the existence of an ocean beneath the ice. The camera looked for active ice volcanism, and craters were counted to allow scientists to estimate the age of various regions. Another goal was the production of a select set of stereo image pairs. Researchers expected to increase the stock of Europa closeups by a factor of seven over the primary mission, with the best images resolved down to just 6 meters! Gravitational anomalies, which could convey information about how thick the ice is in various places, was registered during the flybys, as well, finally, as further data on Europa's magnetic field.

Jupiter's moon Callisto assisted in the final lowering of the perijovium. On four flybys from May 5 to September 16, 1999, Galileo picked up enough momentum to cut its minimum distance from Jupiter in half, to 467,000 kilometers, at the same time, of course, doubling spatial resolution of Jupiter for the camera and NIMS. The Near-Infrared Mapping Spectrometer in particular took advantage of the improved resolution to chart the distribution of water in Jupiter's clouds. Many aspects of the way water circulates vertically, leaving some regions as dry as the Sahara, and the role it plays in weather phenomena remain unsolved puzzles. Meteorologists on Earth may even profit from Galileo's discoveries. During each of the four orbits, the spacecraft shot through Io's plasma torus and should have been able to get a local reading of density values for the sulfur in this ring of charged particles. For Callisto, on the other hand, scarcely any further explorations were planned. The most dramatic phase of the entire Galileo mission, since its arrival in December, was to come finally in the fall of 1999, when the orbiter, under constant and severe bombardment from Jupiter's radiation belt, made two close approaches to Io, one within 500 kilometers and the other within 300 kilometers.

The best images here, like the Europa closeups, were expected to show details down to 6 meters, and Galileo shooting right over the top of the active volcano Pillan Patera's sulfur plume. For good reasons, the return to Io to repeat the closeups missed in 1995 was put off to the very end. The experience from 1995 to 1997 so far, as well as mathematical projections of how severe the radiation will be, had justified hopes that Galileo would keep working all the way through 1999. The hopes for the GEM were largely fulfilled. Radiation damage eventually took its toll in 1998 and 1999, with Galileo botching two of the Europa encounters when the onboard electronics suffered glitches just before data taking was to start. But the underlying problems were identified, critical segments of the control software were changed—and once more Galileo showed it strengths. When it was hit by an unexpectedly large dose of radiation in the vicinity of Callisto in August of 1999, the updated software kept Galileo going. Otherwise, the extended mission posed few risks. The RTGs maintained an adequate power supply for years to come, and fuel supplies are also ample. The tape recorder, however, could fail at any time; it has already been started and stopped in excess of specifications. If it should prove necessary, software that has already been developed will be sent to Galileo, allowing it to transmit a very small amount of data in real time.

Another Mission Extension—and Another One!

As this edition goes to print, the Galileo Europa Mission is already history for more than a year—and the spacecraft is still at work! The public had been kept in the dark about what would happen next as 1999 drew to a close, and then NASA surprised the world with the news of yet another close and successful Europa flyby that had taken place on January 3, 2000. As we have seen, it was the magnetometer data from this encounter that would clinch the case for the ocean of Europa—Galileo's most important observation for some in the community. Only many weeks into 2000 the second extension of Galileo's mission was finally official: it would be called the "Galileo Millenium Mission" and would also include two flybys of Ganymede. Its culmination would come late in 2000, when Galileo and the Saturn-bound Cassini spacecraft would engage in joint studies of Jupiter's vast magnetosphere. It would be the first time in the history of space exploration that two probes would be in the Jovian system at the same time. Cause and effect relationships—how does the magnetosphere respond to changes in the solar wind as well as variations of Io's volcanism?—should become clearer this way.

And then what? There will be a point in time when Galileo's systems will have suffered too much to remain reliable, although the spacecraft has surprised everyone with its robustness, and there are no clear predictions about how likely a major failure is at a given date. The question of how to "dispose of" Galileo became more pressing in the late 1990's, especially in the light of the—however remote—possibility of life in Europa's ocean. By early 2000 it was clear that a crash of Galileo into this moon was to be avoided. The spacecraft had not been sterilized before launch, and some particularly sturdy microbes could still hide somewhere and might even survive a crash into the moon, immediately starting to colonize it and—heavens beware—overrun an indigenous population of "Europans." An extremely remote scenario, all right, but one that cannot be excluded completely. Left alone in its looping orbit around Jupiter, under the influence of the moons' gravity, Galileo could face three different fates: it could crash into the planet itself, into any of the moons or be ejected out of the Jovian system.

Enter NASA's Planetary Protection Officer John D. Rummel. His request for "an assessment of options for the orderly disposal of the Galileo spacecraft at the end of its mission" was answered in the spring of 2000 by the Committee on Planetary and Lunar Exploration (COMPLEX) of the National Academies' Space Studies Board. Its report will be the basis of the decision on Galileo's ultimate fate that NASA Headquarters has to take at some point during 2001. "Obligations imposed by the United Nations' Outer Space Treaty mandate that spacecraft missions be conducted in such a way as to minimize the inadvertent transfer of living organisms from one planetary body to another," the COMPLEX report calls into attention. Simulations of Galileo's orbit indicate that the spacecraft "has a relatively high probability of eventually colliding with one of Jupiter's satellites unless some action is taken to achieve an alternative result. Thus Galileo must be disposed of in a controlled fashion and in a manner that does not compromise the scientific integrity of any planetary body likely to be of interest for future biological studies."

In principle it would be possible to "take advantage of the gravitational interactions between Galileo and Jupiter and its large satellites to engineer a controlled ejection into a heliocentric orbit"—in other words: leaving Jupiter's orbit for good. But such a maneuver would create a tiny but non-zero chance of Galileo eventually returning to Earth! And since the spacecraft is still loaded with nuclear material, another big risk analysis would be necessary, costing in excess of Galileo's annual operations budget of $7 million: "NASA has no option but to dispose of the spacecraft within the Jovian system." This leaves five possible exits: either one of the four Galilean moons or Jupiter itself. Only an impact into the planet or into Io would raise no planetary protection issues, the COMPLEX report concludes. Both bodies are extremely hostile to all life (as we know it, that is), and in the case of Jupiter the atmospheric probe hadn't been sterilized, as it was to be thoroughly destroyed soon after making contact.

Impacts into Ganymede, Callisto or especially Europa, however, should be avoided, as all these moons could carry, in principle, traces of past biological activity or even possess oceans today. Europa is even called "one of the places in the Solar System with the greatest potential for the existence of life" in the report, while it is currently impossible to "quantify the exact locations of Ganymede and Callisto on the Io-Europa spectrum of planetary-protection concerns." Galileo's mission planners favor, for operational reasons, a controlled impact into

Jupiter as the best way out, and COMPLEX agrees with that assessment. Just before this book went to press, NASA made its final decision on the future, and end, of the Galileo mission, following the suggestions of Planetary Protection Officer Rummel and the COMPLEX report. Galileo's last mission extension includes five more flybys of the Jovian moons before being sent on its final plunge into the crushing pressure of Jupiter's atmosphere.

On May 25, 2001, Galileo should have passed about 123 kilometers above Callisto, the second largest of Jupiter's 28 known moons. The effects of Callisto's gravity will set up the space probe for a swing over both polar regions of the intensely volcanic moon Io in August and October 2001. During these two flybys, Galileo will take pictures, measure magnetic forces, and study dust and smaller particles. Science goals include studying the extent of volcanism on Io, both in new and previously active sites, determining whether Io generates its own weak magnetic field, and gaining a better understanding of the moon's doughnut-shaped ring, the Io Torus, that encircles Jupiter and contains electrically charged gases. In 2002, having completed its imaging mission, Galileo will continue studies of Jupiter's massive magnetic field with seven instruments. In January, the orbiter will fly near the equator of Io.

In November 2002, Galileo will swing closer to Jupiter than ever before, dipping within about 400 kilometers of the moon Amalthea, which is less than one-tenth the size of Io and less than half as far from Jupiter. Scientists will use Galileo's measurements to determine the small moon's mass and density. They will also study dust particles as Galileo flies through Jupiter's gossamer rings and seek new details of the magnetic forces and the densitites of charged particles close to the planet.

Galileo's final orbit will take an elongated loop away from Jupiter. Then, in August 2003, the spacecraft will head back for a direct impact and burn up as it plows into Jupiter's 60,000-kilometer-thick atmosphere.

During Galileo's third mission extension, the probe's health status should be monitored even more closely than before. "The detection of loss in redundancy in any critical command and control subsystem should trigger the initiation of the appropriate maneuvers necessary to place Galileo on a ballistic trajectory designed so that the spacecraft will collide with Jupiter," stated the COMPLEX report of 2000. Will there be some science possible during that ultimate "disposal"? Not while Galileo is actually plunging into Jupiter's atmosphere and quickly burning up, says former mission manager Jim Erickson. There is a delay of several minutes inside the spacecraft between taking data and having them ready to broadcast, and the data rate is so low that hardly anything could be sent to Earth in time. But perhaps some lower and lower passes over Jupiter's clouds will be possible and remote sensing and fields and particles data be collected.

As of October 2000 Galileo's health was "pretty good," reports Erickson. All problems within the spacecraft's subsystems that had shown up over the years have been worked around, and all of the scientific instruments—apart from the UV Spectrometer that doesn't produce useful data anymore—are operational as well. And the tape recorder that had caused so much alarm in 1995? It has behaved surprisingly well in the following years, with no new surprises. Best of all, during the first five years in the Jovian orbit there has not been one close call where some malfunction almost caused a loss of the whole mission. Such an experience, not uncommon in planetary exploration, was spared the Galileo team. The staff working on Galileo's day to day operations is now down from several hundred during the primary mission to about 45—some of them veterans from the early days, all with at least two years of experience. The annual operations budget is now around $12 million, a bit lower than the $15 million during the

Galileo Europa Mission and way smaller than the $55 million that it took each year to run the primary mission. Few of the scientists who were with the mission from the beginning have left the project, by the way—a sign that it is still worth continuing the effort.

What are the lessons from Galileo for future missions to the outer planets? Erickson is most amazed that Galileo has proven so "bullet-proof" to the harsh radiation environment of Jupiter. Already the spacecraft has experienced three times the radiation dose it was designed for, and yet the number of electronics glitches was surprisingly low. Perhaps Jupiter was a bit "milder" than the planners had thought—or Galileo is just a very well designed spaceship. And it was also a rather pleasant one to operate: the unique way the mission has been run since 1996, with one week of intense action followed by two months of cruise, that the need to download the tape recorder very slowly has forced upon the flight team, was actually a blessing in disguise. People could go on vacation and not miss much and be back for the next flyby of a moon. The lesson here is: build enough down-time into your mission plan if you want to keep the team alert for years on end. When it is all over, some 14 years will have passed since Galileo's launch and over 30 years since the mission was conceived. For a long time books and articles will go on being written about Galileo's discoveries and the puzzles that remain. Perhaps a later generation of researchers will even come across phenomena in the data that were overlooked in the rush of mission operations. Discoveries are still being made from the Voyager pictures, for example, over a decade after they were made.

Beyond Galileo: Keeping Our Sights on Europa

Above all, however, the breathtaking closeups of Europa inspired a wealth of ideas for new missions to this promising world—and the first has already won approval! The NASA budget for the 1999 fiscal year included the first funds for the development of the Europa Orbiter, which was supposed to become the first *economy* excursion to the outer Solar System. Initial plans called for a launch date in 2003, then a few years under way, before scientists will use the orbiter's radar to systematically measure the thickness of Europa's ice crust. Because ice is deeply permeable to the long radio waves, they can be used to find out whether any distinct boundary exists between frozen and liquid water—perhaps as little as 1 kilometer beneath the surface. At the same time, other onboard instruments will be making detailed maps of the ice crust, which will enable researchers to answer questions about what is going on inside this fascinating moon. As long as they do not cost too much for the limited budget, the orbiter might even be equipped with small remote seismometers that it could shoot into the ice, where they would lodge and transmit data.

The new NASA program for exploring the outer planets ran into severe budgetary as well as technological difficulties in 2000, however, and the launch date of Europa Orbiter has already slipped to 2008, the earliest—no data from Europa can be expected until well into the following decade. A companion mission to the planet Pluto, the Pluto Kuiper Express, which was to go a few years before the Europa Orbiter and to test some of its new systems, has been issued a stop-work order in September 2000, after the combined cost of both missions had almost doubled to $1.3 billion from the initial estimate of $700 to 800 million—so much for exploring the world beyond the Asteroid Belt on the cheap. Now most of the Pluto people from JLP will be drafted into the Europa Orbiter development team, while NASA has launched an open competition in December 2000 for a possibly less expensive Pluto mission. The whole Outer Planets program—which also includes a daring mission close to the Sun that would share many technologies—is meanwhile undergoing major re-structuring, but to many of NASA's key scientists the Europa Orbiter remains the most important mission in this context. The hope of bolstering the case for possible life in Europa's ocean has become a major driver for programmatic decisions.

The Europa Orbiter, however, was only one of many proposals for how we might learn more about the mysterious moon. Another idea was the "Europa Ice Clipper." Designed to hurl a 10- to 20-kilogram projectile at a speed of 10 kilometers per second at the moon, casting up a storm of ice, it would use aerogel—a kind of frozen foam—to gently capture samples of the ice and keep them frozen for delivery to Earth. The Clipper was practically a copy of the Stardust probe, which had already been developed for a 1999 launch toward the comet Wild 2, where it will use the same aerogel technology to collect dust particles and bring them to Earth. Presumably, it would have been even cheaper to build than the Orbiter, but it could not have undertaken any systematic investigation of Europa as a whole. And that would be indispensable for the next really big step—giving a definitive answer to the question of whether there really is liquid water under the ice and, if so, preparing the way to explore the subsurface ocean.

The high point in our exploration of Europa is supposed to be a soft landing on one of the thinner parts of the ice crust, six years or so into the mission, after the Orbiter has delivered a detailed three-dimensional image of the moon's structure and composition. Just how the rest of the mission should go has been under consideration for several years already. A drill, called a "cryobot," is supposed to bore through the ice to release a tiny submarine (a hydrobot) into the ocean, which will be designed to transmit data along a cable to the lander on the moon's surface, for broadcast back to Earth. The technology for this bold venture is being tested in the Antarctic, as part of an expedition to

Lake Vostok, which for millions of years has been covered by a sheet of ice 4 kilometers thick. At the announcement of the Europa Orbiter in February 1998, much was made of the point that an explicit NASA goal is the development of a submarine capable of operating in alien worlds.

Operations in the Jovian system pose a far greater challenge than any excursion to Mars. The distance is immense, the solar energy weak—and the budget is limited. New interplanetary probes, except those involved in the systematic exploration of Mars, were possible up until 1998 in the U.S. only as part of the "Discovery Program," which imposes strict cost limitations, in part by calling for lightweight probes capable of being launched with small, economical rockets. In principle, the entire Solar System is fair game for a Discovery probe, but only missions with "simple" destinations have been accepted so far, none any farther away than Mars. NEAR (Near Earth Asteroid Rendezvous) was sent to visit a pair of asteroids, and Pathfinder is on its way to Mars without having to face any competition. One Moon orbiter, two comet missions, and a probe to collect samples of the solar wind have won selection over rivals and were next in line.

Asteroid enthusiasts, then, who have always had to hitch a ride on other, already approved missions, are paying their own way—but what about the rest? Critics bemoaned a discouraging trend toward the exclusive approval of low-cost proposals for missions to exotic destinations at the expense of continuing the systematic exploration of places we have already been. The result is an ever-growing list of new questions, rather than answers to a coherent set of questions that we already know. On the other hand, it also makes sense to launch limited-payload Discovery missions to destinations that have not been researched at all, where one or two instruments are capable of delivering fundamental new insights.

The Europa orbiter was meant to inaugurate a new series of spacecraft—the "Outer Planets/Solar Probe line"—which were once hoped to be as economical as the others, but can range anywhere in the Solar System. Part of this category, in addition to the Europa Orbiter, is the "Solar Probe," designed to come extremely close to the Sun, as well as the now abandoned "Pluto Kuiper Express," a mission to the planet Pluto and other small celestial bodies on the edge of the Solar System. The Solar Probe, the Pluto Express, and, indeed, the Europa Orbiter had once been promoted together as the "Fire and Ice" project, because many of the technological problems they pose are supposed to be identical. Although the primary objectives of all three missions differ sharply, they could conceivably share the same electronics, the same software, and even the same flight controllers. By combining aspects, the cost of each "Fire and Ice" mission could perhaps be kept below that of a Mars Pathfinder—a prospect that just a few years ago would have sounded like science fiction (and perhaps will remain just that, for the forseeable future). How much payload will actually be able to go along, and how such missions will compare in terms of scientific gain with the large-scale endeavors of the past, only the future will tell.

Or were the expensive "old" missions perhaps already not the economical solution from the point of view of optimizing science? Advantages and disadvantages tend to balance out. The old approach, getting as many multifaceted experiments as possible to a distant location and exploring what was there with the maximum possible number of techniques simultaneously, does have something to say for itself in terms of science. The disadvantages, however, are also considerable. Flexibility suffers and many compromises are required to try to get the most from every instrument. And the cost of each mission runs into the billions of dollars. Finally, all the risk falls on a single mission. If anything goes wrong, either at the launch or with the spacecraft itself, dozens of experiments are lost at once.

No less considerable are the disadvantages inherent in the new generation of smaller, less risky probes like the Mars Pathfinder or the Mars Global Surveyor. The systematic exploration of alien worlds with a dozen different instruments at once is a thing of the past; the devices arrive at their goals in succession, if at all. The scientific value of the Mars Pathfinder is questioned by some critics. The approach requires all of the subsystems making up the spacecraft, from fuel supply to onboard electronics, to be built anew each time. Even the "cheap" missions run in the neighborhood of $100 to $300 million. To accumulate all of the successes of the Galileo mission from 1989 to 2000 (and beyond) with a series of Discovery probes—reproducing all of the planetary and asteroid flybys, witnessing the crash of the Shoemaker-Levy comet into Jupiter, the many close approaches to Jupiter's moons—would have run into the billions of dollars, if it even proved possible. The last, biggest, and most expensive of all the dying class of giant probes, in any case, is Cassini, which has been speeding toward Jupiter's sister planet Saturn since October 1997.

The Future Has Begun: Cassini on the Way to Saturn

The Cassini launch on October 15, 1997, came off extraordinarily well. The Centaur main stage of the Titan 4B rocket functioned so precisely that flight controllers were able to reduce the first course correction on November 9 to a minimum. The fuel saved will come in handy later on in Cassini's complicated journey—possibly increasing the scientific yield of the mission. At first it had seemed that Cassini would have to spend the seven-year "cruise" between Earth and Saturn basically asleep. No money had been found to do scientific observations *en route,* especially during the flybys of other planets. But every so often NASA's can-do attitude prevails, and indeed this is what happened.

Jupiter viewed through three Cassini filters that sample three wavelengths where methane gas absorbs light: in the red at 619 nanometer wavelength (top), and in the near-infrared at 727 nanometers (middle) and 890 nanometers (bottom). Absorption in the 619 nanometer filter is weak, stronger in the 727 nanometer band and very strong in the 890 nanometer band, where 90 percent of the light is absorbed by methane gas. Light in the weakest band can penetrate the deepest into Jupiter's atmosphere and is sensitive to the amount of cloud and haze down to the pressure of the water cloud (about six times the atmospheric pressure at sea level on the Earth). Light in the strongest methane band is absorbed at high altitude and is sensitive only to the ammonia cloud level and higher (pressures less than 1/2 of an atmosphere), and in the middle methane band it is sensitive to the ammonia and ammonium hydrosulfide cloud layers as deep as twice the atmospheric pressure. The images shown here demonstrate the power of these filters in studies of cloud stratigraphy. The most prominent feature seen in all three filters is the polar stratospheric haze, which makes Jupiter bright near the pole.

More and more mid-cruise science was enabled while the mission proceeded. When Cassini arrived at Jupiter on time for the turning of the millennium, a full-blown research program in the style of the Voyager encounters was possible.

This was a nice bonus (plus a good trial run for Saturn) for a mission with a huge pricetag—the Cassini adventure is costing $3.4 billion, an incredible sum for a single, unmanned space flight. Nevertheless, from 1990 to 2008, NASA and its junior partners from the European Space Agency will have spent something approaching this total on the last great planetary mission, which involves the systematic exploration of Saturn and its rings, moons, magnetic field, and all the charged particles trapped inside it. The mission is supposed to continue for at least four years after Cassini's arrival. It is in principle a repeat of the Galileo mission for the other giant planet in the Solar System, but it will be even more complex and take advantage of more advanced instrumentation. A descent probe makes up an important part of this mission as well, but Huygens, as the probe is called, rather than plunging into Saturn's atmosphere, will make a soft landing on Titan, its biggest moon.

The Cassini program got under way in the 1980s, when the exploration of interplanetary space was being conducted with ever larger and more complicated probes. Recalling the Galileo experience, the enormous costs brought Cassini face to face with cancellation on more than one occasion, requiring drastic reductions in 1992. Among the necessary sacrifices was the moveable camera platform designed on the Voyager and Galileo model. And even so, full of fuel and mounted with the Titan-probe Huygens, Cassini came in at about 5.65 tons in weight, 6.8 meters in height, and 4 meters across—following in this regard the massive Russian Mars probe, the biggest object ever sent to another planet.

Nevertheless, not even the Titan-4B rocket and the most powerful Centaur main stage (loaded with 22 tons of fuel) were able to send

These three images, taken through the narrow angle camera from a distance of 77.6 million kilometers on October 8, 2000, reveal more than is apparent to the naked eye through a telescope. The image on the left was taken through the blue filter (centered at 445 nanometers in wavelength), within the part of the electromagnetic spectrum detectable by the human eye. The appearance of Jupiter in this image is, consequently, very familiar; the Great Red Spot and the planet's well-known banded cloud lanes are obvious. Jupiter's appearance changes dramatically in the images to the right, which were taken in the ultraviolet at 255 nanometers (middle) and in the near infrared at 889 nanometers (right). These images are near negatives of each other and illustrate the way in which observations in different wavelength regions can reveal different physical regimes on the planet. All gases scatter sunlight efficiently at short wavelengths, an effect that is more pronounced in the ultraviolet. The deep-banded cloud layers are thus made invisible in the middle image, and only the very high altitude haze appears dark against the bright background. The contrast is reversed in the near infrared, where the abundant methane gas is strongly absorbing and therefore appears dark. Again the deep clouds are invisible, but now the high altitude haze appears relatively bright against the dark background. High altitude haze is seen over the poles and the equator. The Great Red Spot, prominent in all images, is obviously a feature whose influence extends high in the atmosphere.

Cassini on the most direct route to Saturn. Instead, the spacecraft will loop from what seems to be one side of the Solar System to the other, taking over 6 1/2 years to reach its destination. Just like Galileo, Cassini had to pick up more than one gravity assist on the way—twice from Venus and one each from Earth and Jupiter. So the trip to Saturn also began with a detour to Venus. The first flyby took place in April

1998 and the second in June 1999. The return trip for a close Earth flyby took only 56 days, and on December 30, 2000, Cassini reached its last target *en route* to Saturn, the giant planet Jupiter.

While the central purpose of the Jupiter flyby had always been to gather speed and somewhat shorten the trip to Saturn, NASA just couldn't forgo the chance for a big trial run of real science operations at a real giant planet, although the necessary funding had been secured only well after Cassini's launch. For certain observations Cassini's suite of instruments was better suited than the Voyagers' or Galileo's—and there was the added bonus of simultaneous observations by Galileo and Cassini! In fact, these joint observations, spanning five months (October 2000–March 2001), were the first time in the history of deep space exploration that spacecraft from two independent robotic missions would actively and cooperatively observe the same giant gas planet at the same time from close range. Cassini was expected to remain outside Jupiter's magnetosphere most of the period, whereas Galileo would transition between Jupiter's magnetosphere and the solar wind, depending on where it was in its orbit. This provided scientists with the opportunity to study the solar wind and its effects from both within and without the magnetosphere—a major step forward, given the complexities of Jupiter's magnetic effects.

When Cassini's campaign began in early October 2000, good and bad news was released on both sides of the Atlantic almost simultaneously. While NASA was hailing the first pictures taken from 84 million kilometers away ("clearly shows the exceptional resolving power of the imaging system"), ESA had to disclose a major design error in the Huygens capsule—or rather in the receiver for its signals aboard the Cassini mothership, which had also been built in Europe. Only in 2000 had tests shown that the bandwidth of this receiver was too small to cope with the shift in Huygens's frequency due to its varying speed

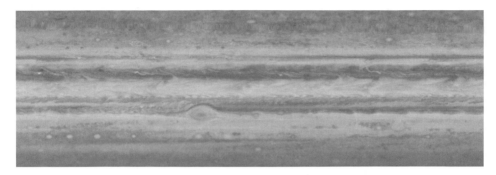

A full view of the Jovian atmosphere, between 60° south and 60° north. Maps like this one (part of a long movie sequence covering the first week of November 2000) are made by first creating a mosaic of six narrow angle images taken in the same spectral filter over the course of one Jupiter rotation and, consequently, covering the whole planet. Three such global maps—in red, green, and blue filters—are composited to make one color map, showing Jupiter during one Jovian rotation. The smallest visible features at the equator are about 600 kilometers.

relative to Cassini. This oversight was a major scandal, and ESA immediately started several investigations: Who had screwed up? (This was never found out, due to the complicated structure of the project.) Were other ESA spacecraft affected? And what could be done to rescue Huygens's unique mission?

While Cassini was edging ever closer to Jupiter, with the resolution of the pictures eventually increasing tenfold (the flyby distance was 9.7 million kilometers) and data from other instruments coming in as well, ESA feverishly looked for answers. A report released on December 20, 2000, gave some hope. There would be ways to change the geometry between Cassini and Huygens during the critical hours of the communications link, probably without wasting a lot of Cassini's precious maneuvering fuel. All data broadcast during Huygens's descent in Titan's atmosphere would thus be recoverable. A detailed strategy would have to be worked out between NASA and ESA during

2001, but in principle the snafu could be overcome. Still it was troubling to read that even in a multi-billion dollar project the same basic mistakes could happen as in the comparably cheap NASA Mars mission disasters of 1999. If just the right questions would have been asked before Cassini's launch, a simple technical solution would have been possible in no time. . . .

With the Huygens scare just fading, a sudden technical glitch struck Cassini. When Cassini needs to reorient itself in space, it normally doesn't use its thrusters, but electrically driven reaction wheels. When they are moved in one direction, the spacecraft rotates in the opposite one. During the start-up of one of the reaction wheels in mid-December, 2000, unexpected friction was experienced, and all of the science operations that required pointing Cassini had to be stopped. A series of tests quickly determined that the friction only appeared in one particular wheel, and only when it was rotating slowly: the lubricant seemed to behave strangely and to require high rotation rates. By isolating this malfunction, scientists were able to use the reaction wheels again, and the pointing observations resumed two days before the closest approach. It was quite fortunate that the problem had arisen already during the Jupiter campaign. When another lubrication problem had struck Voyager 2 during its only Saturn flyby in 1981, a lot of valuable science had been lost. The Jovian trial run had already paid for Cassini.

Apart from the reaction wheel incident, there were only happy faces in the Cassini team. "The spacecraft is steadier than any spacecraft I've ever seen," Carolyn Porco, team leader for the camera on Cassini, said after seeing the first images in October: "It's so steady, the images are unexpectedly sharp and clear, even in the longest exposures taken and most challenging spectral regions." Soon the first color images were created by using different filters, and then sequences of pictures were combined into little movies. Had Galileo's main antenna

This image, made from narrow angle images taken on December 12, 2000, captures the innermost Galilean satellite, Io, and its shadow in transit against the disk of Jupiter. The distance of the spacecraft from Jupiter was 19.5 million kilometers; the spacecraft latitude was 3° above Jupiter's equator plane. The image scale is 117 kilometers per pixel.

worked, such movies would have been delivered all the time, but now Cassini finally closed this major gap. A few hours after the closest approach, the scientists were ready for a first summary of discoveries— and there were a lot to be discussed. One major finding was about the large storms on Jupiter, which can be larger than Earth and last for centuries by gaining energy from swallowing smaller storms. These smaller storms pull their energy from lower depths.

Cassini's pictures of thunderstorms on Jupiter—and especially the movie sequences—were the key: as small storms pass each other, they can be ripped apart or merged. This shows that the small features in Jupiter's atmosphere harvest the energy from below the cloud surface, and the larger storms encompass the small ones, just as a big fish eats smaller ones for energy (this concept is actually known as the "fish model" among researchers). Before the Cassini movies, the fish model had been one of several, but now it seems to have won. A better understanding of the storms on Jupiter also helps in understanding Earth's atmosphere, where the weather is very different from Jupiter's. Why is Jupiter's weather so stable, with some storms like the Great Red Spot lasting for 300 or more years, while the weather on Earth is so transient?

Other data from Cassini dealt with the magnetosphere, and here the joint observations with Galileo already had paid off. Both spacecraft have returned evidence of the variability in size of Jupiter's magnetosphere, the bubble of charged particles trapped within Jupiter's magnetic field. The bubble is so big that if it were visible to the eye, it would appear bigger to viewers on Earth than our own Moon, despite its much greater distance. While Galileo was moving toward Jupiter in the fall of 2000, it passed the magnetosphere boundary, but then the boundary moved inward toward Jupiter even faster than the spacecraft was moving, temporarily putting Galileo back outside the magnetosphere.

And there is more. Another instrument on Cassini, the Magnetospheric Imaging Instrument (MIMI), is creating images never before possible. It's not a camera in any classical sense, but by tracing back the trajectories of particles racing away from the Jupiter system, MIMI can image the entire magnetosphere, so to speak. Other Cassini measurements show that some sulfur and oxygen spewed from volcanos on Io are distributed much farther from Jupiter than its extent of the

The Galilean satellites Europa and Callisto are caught, under the watchful gaze of Jupiter, nearly perfectly aligned with each other and the planet's center in this image taken on December 7, 2000. The distances here are deceiving. Europa (seen against Jupiter) is 600,000 kilometers above the planet's cloud tops; Callisto (at lower left) is nearly three times that distance at 1.8 million kilometers.

magnetosphere. The evidence shows that there is a big nebula of material surrounding Jupiter, originating from the volcanos on Io and traveling with the planet around the Sun. But Cassini is definitively going
the other way. A major braking maneuver on July 1, 2004, will put the
spacecraft into orbit around Saturn. That moment, as it did for Galileo
and Jupiter, marks the actual start of the Cassini mission, which is
scheduled to continue until at least 2008. Only once the last broadcast
signals from Cassini finally reach Earth will the end come for the second era in the exploration of the outer Solar System, which began two
decades earlier with the launch of Galileo.

Timeline

1969		Planning starts for an Outer Planets Grand Tour (OPGT), inaugurating the Pioneer program
1972	March 2	*Pioneer 10 launch*
	July 1	Voyager program gets underway
1973	April 5	*Pioneer 11 launch*
	December 3	Pioneer 10 Jupiter flyby
1974		First concrete plans made for a Jupiter orbiter

	December 2	*Pioneer 11 Jupiter flyby*
1976		Original idea for a Jupiter orbiter-and-atmospheric capsule
1977	July 1	Galileo program gets underway
	August 20	*Voyager 2 launch*
	September 5	*Voyager 1 launch*
1979	March 5	*Voyager 1 Jupiter flyby*
	March 8	Discovery of the first active volcano on Io
	July 9	*Voyager 2 Jupiter flyby*
1983		Galileo launch set for May 1986
1986	January 28	*Challenger disaster brings the space program to a standstill*
	June	Galileo loses its Centaur main stage rocket
1987	February 21	Galileo is back at JPL
	December	VEEGA trajectory for Galileo decided on
1988	May 16	Galileo arrives at Kennedy Space Center for the second time
1989	end of August	Atlantis and then Galileo are put on the launch pad
	mid-September	White House OK's the launch
	October 10	Judge Gasch rejects petition to stop the launch
	October 18	**Galileo lifts off**
	November 9–11	Galileo makes first course correction
1990	February 9	**Venus flyby**
	November	Playback of majority of Venus data to Earth
	December 8	**First return to Earth**
1991	March 20	First of several course corrections for approach to the asteroid Gaspra

	April 11	**High-gain antenna fails to open**
1991	May 20–23	Warming up the antenna does not work
	July–August	Freezing the antenna fails to work twice
	October 29	**Asteroid Gaspra flyby**
	November 15 to April 1992	Galileo's great distance from Earth forces data rate down from 40 to 10 bits per second
	from mid-December	Galileo allowed to get even colder to free the antenna—fails to work
1992	from January	Renewed warming, other attempts fail to release antenna
	January 13–30	Galileo is almost exactly behind the Sun, disturbing radio contact (which is also put to scientific use)
	February 8	*Ulysses picks up a gravity assist from Jupiter to escape the ecliptic plane*
	April 29	Galileo's antenna motor switched on momentarily for diagnostic purposes—nothing moves
	May—June	JPL presents first closeups of Gaspra
	June	NASA announces attempt to "hammer" the high-gain antenna open during next Earth flyby; should that fail, planning will start in March 1993 for exclusive reliance on the low-gain antenna
	August 4–8	First course correction in preparation for second Earth flyby
	December 8	**Second return to Earth**
	end of December	Discovery of a magnetic field on Gaspra announced

	December 28–	Attempt to hammer antenna open fails; all
	January 19, 1993	efforts focused now on using the low-gain antenna to salvage the mission
1993	March 21–April 11	Galileo, Ulysses, and the Mars Observer search for gravitational waves
	May 22	Comet crash of P/Shoemaker-Levy 9 first predicted
	August 28	**Asteroid Ida flyby**
	September 22	First photograph of Ida released
	early October	Major 5–day course correction puts Galileo into final Jupiter trajectory
	October 21	Results of Earth flyby analysis published in *Nature*
1994	March 3	Discovery of Ida's moon announced
	July 16–21	**Crash of SL-9 into Jupiter observed directly by Galileo**; data playback till January 1995
1995	February	Galileo receives a new operating system
	March 13	New software takes over flight control
	July 7	Atmospheric probe activated
	July 13	Release of atmospheric probe
	July 27	First firing of Galileo's main engine
	August	Great Jovian "dust storm"—up to 20,000 hits a day lasting into fall
	October 11	Galileo photographs Jupiter on approach; tape recorder fails on playback
	December 7	**Arrival day at Jupiter**: Io passage; data reception from atmospheric probe; Jupiter orbit insertion
	December 10	First playback of probe data

1996	January 22	Major press conference announcing very preliminary results
	March 14	Third main engine burn raises the perijovium out of the radiation zone
	March 25	Absolutely final—and fruitless—attempt to hammer open the high-gain antenna
	April 15	Complete playback of atmospheric probe data, despite continuing snags with the tape recorder
	May and June	Loading new software, this time for the orbital tour
	May 10	Publication of relatively reliable numbers from the atmospheric probe
	June 27	**First Ganymede flyby** at 835 km distance (G1)
	September 6	Second Ganymede flyby at 262 km distance (G2)
	November 4	First Callisto flyby at 1,100 km distance (C3)
	December 19	First Europa flyby at 695 km distance (E4)
1997	February 20	Second Europa flyby at 588 km distance (E6)
	April 5	Third Ganymede flyby at 3,065 km distance (G7)
	May 7	Fourth Ganymede flyby at 1,584 km distance (G8)
	June 25	Second Callisto flyby at 416 km distance (C9)
	September 17	Third Callisto flyby at 524 km distance (C10)
	November 6	Third Europa flyby at 1,119 km distance (E11)
	December 7	**Start of Galileo Europa Mission (GEM)**

	December 8	Bob Mitchell replaces Bill O'Neil as Galileo project manager (to be followed by Jim Erickson and Eilene Theilig)
	December 16	First and closest GEM Europa flyby at 200 km distance (G12); partial failure of Galileo's attitude control system; data transmission rate suffers
1998	January 13	Antenna orientation mostly back under control
	February 10	Europa flyby at 3,552 km distance
	March 29	Europa flyby at 1,649 km distance
	May 31	Europa flyby at 2,521 km distance
	July 21	Europa flyby at 1,837 km distance
	September 26	Europa flyby at 3,582 km distance
	November 22	Europa flyby at 2,273 km distance
1999	February 1	Europa flyby at 1,495 km distance
	May 5	Callisto flyby at 1,311 km distance
	June 30	Callisto flyby at 1,048 km distance
	August 14	Callisto flyby at 2,299 km distance
	September 16	Callisto flyby at 1,052 km distance
	October 11	Io flyby at 611 km distance
	November 26	Io flyby at 300 km distance
2000	January 3	Europa flyby at 199 km distance, inaugurating the **Galileo Millenium Mission**
	February 22	Io flyby at 199 km distance
	May 20	Ganymede flyby at 809 km distance
	June	Galileo leaves Jupiter's magnetosphere for the first time since 1996
	September 8	Apojove (290 Jupiter radii = 20.7 million km) during the largest orbit since arrival
	December 28	Ganymede flyby at 2,321 km distance

Keep Up to Date!

Books, Magazines, and Internet Addresses

Magazines and periodicals

that frequently report on the progress of the mission and its findings

The Planetary Report (6 times yearly, American)
   c/o The Planetary Society, 65 North Catalina Avenue,
   Pasadena, CA 91106-2301
   On the Internet: http://planetary.org
Sky & Telescope (12 times yearly, American)
   c/o Sky Publishing Corporation, P. O. Box 9111,
   Belmont, MA 02178-9111
   On the Internet: http://www.skypub.com

Aviation Week & Space Technology (about 50 times yearly,
American)
c/o McGraw-Hill, 1221 Avenue of the Americas,
New York, NY 10020
On the Internet: http://www.aviationnow.com

Galileo and Planetary Science on the Internet

All links can be accessed directly via:
http://www.geocities.com/skyweek/galileo

Galileo

The Galileo project home page is at: http://www.jpl.nasa.gov/galileo
All of Galileo's Jupiter mission photographs are at:
http://www.jpl.nasa.gov/galileo/images.html
Galileo's science data archive is at:
http://nssdc.gsfc.nasa.gov/planetary/galileo.html

Cassini, Voyager, and Pioneer

Cassini's Jupiter Flyby
http://www.jpl.nasa.gov/jupiterflyby
Jupiter Imaging Diary
http://ciclops.lpl.arizona.edu/ciclops/images_jupiter.html
The Voyager project home page, created after the fact, is located at:
http://vraptor.jpl.nasa.gov
The ongoing Voyager Interstellar Mission "lives" at:
http://vraptor.nasa.gov/voyager/vimdesc.html
And information about Pioneer 10 and 11 are at:
http://pyroeis.arc.nasa.gov/pioneer/PN10&11.html

The Planets on the Internet

Selected photographs from NASA planetary missions are available at:
http://photojournal.jpl.nasa.gov

"Views of the Solar System" are at:

http://www.solarviews.com/eng/homepage.htm

"Welcome to the Planets," says the Planetary Data System at:

http://pds.jpl.nasa.gov/planets

"The Nine Planets Multimedia Tour" begins at:

http://www.seds.org/nineplanets/nineplanets

And an introduction to the moons of the outer planets can be had at:

http://css.jsc.nasa.gov/pub/research/outerp/moons.html

News Reports about Astronomy and Space Travel

http://www.flatoday.com/news/space

http://www.spaceflightnow.com

http://www.space.com

http://www.spaceref.com

http://www.geocities.com/skyweek/mirror

Books about the Planetary System

Brief annotated selection

J. K. Beatty *et al.*, eds., *The New Solar System,* 4th edition (New York: Cambridge University Press, 1998). Particularly in-depth essays about aspects of the planetary system from the pens of active researchers.

D. Fischer, H. Heusler, *Der Jupiter Crash,* 2nd updated edition (Birkhäuser 1996). Still the only protocol of the great comet crash of 1994 in German; now with detailed analyses.

L. Ksanfomality, *Planets* (Moscow: MIR, 1985). Detailed description of the planetary system from the perspective of a leading Soviet researcher.

K. R. Lang, C. A. Whitney, *Wanderers in Space: Exploration and Discovery in the Solar System* (New York: Cambridge University Press, 1991). Solid, systematic introduction to all aspects of the planetary system.

P. Moore, G. Hunt, *Atlas of the Solar System,* 2nd edition (Chicago:
Rand McNally, 1984). Comprehensive compilation of mono-
graphs on the individual planets and the Sun.

D. Morrison, J. Samz, *Voyage to Jupiter* (Washington: NASA SP-439,
1980). The book to own about the Voyager mission, and possibly
about NASA.

I. Peterson, *Newton's Clock: Chaos in the Solar System,* reissue edition
(New York: W. H. Freeman, 1995). In addition to the description
of chaos in the Solar System, also a very readable introduction to
celestial mechanics.

Planeten und Monde (Heidelberg: Verlag Spektrum der Wissenschaft,
1988). A collection of 17 *Spektrum* magazine articles from the
1980s.

Jupiter Research in Periodicals

(Among the periodicals listed—to be found in university libraries or
the libraries of observatories—only *Sterne und Weltraum* is in German; all others in English.)

Popular Overviews

Gehrels, *Sky and Telescope,* 2/1974, 76, and Watts, ibid., 79.
Pioneer 10 at Jupiter!

NN, *Sky and Telescope,* 2/1975, 72: Pioneer 11 at Jupiter

Köhler, *Sterne und Weltraum,* **16**, 316 (10/1977): The Voyager project

Beatty, *Sky and Telescope,* 5/1979, 423, 6/1979, 516, and 9/1979, 206,
as well as Köhler, *Sterne und Weltraum,* **19**, 16 (1/1980): The two
Voyager probes at Jupiter

Johnson and Yeates, *Sky and Telescope,* 8/1983, 99: Comprehensive
descriptions of the reborn Galileo project

Carroll, *Sky and Telescope,* 4/1987, 359: New comprehensive description of Galileo in the post-Challenger world

Fischer, *Sterne und Weltraum,* **28**, 714 (12/1989): Comprehensive treatment of the mission after the successful launch

Beatty, *Sky and Telescope,* 3/1991, 269: Galileo's 1990s observations of Venus, Earth, and the Moon

Neukum and Oberst, *Sterne und Weltraum,* **30**, 307 (5/1991): Details of the Moon observations

Fischer, *Sterne und Weltraum,* **31**, 103 (2/1992) and Beatty, *Sky and Telescope* 2/1992, 134, as well as 9/1992, 253: Galileo's Gaspra findings

Beatty, *Sky and Telescope* 4/1993, 18: Halfway to Jupiter!

Fischer, *Sterne und Weltraum,* **32**, 339 (5/1993): The second Earth flyby and the end of the main antenna

Fischer, *Sterne und Weltraum,* **34**, 520 (7/1995): Interview with the Galileo head scientist

Fischer, *Sterne und Weltraum,* **34**, 712 (10/1995): Galileo's contribution to understanding the course of the comet crash

Doody, *Sky and Telescope,* 12/1995, 30, and Johnson, *Scientific American,* 12/1995, 28: Galileo is finally there!

Fischer, *Sterne und Weltraum,* **35**, 274 (4/1996): and Beatty, *Sky and Telescope,* 4/1996, 20: First discoveries of the Galileo probe

Chapman, *Astronomy* 6/1996, 47: What did we learn from Gaspra and Ida?

Fischer, *Sterne und Weltraum,* **35**, 720 (5/1996): Discoveries of the first Ganymede flyby

Fischer, *Sterne und Weltraum,* **35**, 816 (5/1997): New indications of an ocean on Europa

Spencer, *New Scientist,* April 5, 1997, 42: Galileo's discoveries on the Jovian moons

Millstein, *Astronomy,* October 1997, 38: What we know about Europa's ocean (and future research questions)

Carroll, *Sky and Telescope,* 12/1997, 50: Life in an ocean on Europa? (and future research questions)

Denk & Althaus, *Sterne und Weltraum,* **37**, 18 (1/1998): Detailed description of Galileo's Europa mission

Semipopular Overviews

Baguhl & Grün, *Sterne und Weltraum,* **32**, 669 (7/1993): Ulysses dis-
covers Jupiter's dust streams

Reichardt, *Nature,* **377**, 669 (October 26, 1995): What would be the
significance of the tape recorder failing?

Ingersoll, *Nature* **378**, 562 (December 7, 1995): Speculation about
what it is like inside Jupiter

Cowen, *Science News,* **149**, 55 (January 27, 1996), Carlowicz, *Eos,* **77**
(January 30, 1996), #5, and Kerr, *Science,* **271**, 593 (February 2,
1996): Confusing observations by the Galileo probe

Levy, J., *Royal Astronomical Society of Canada,* **90**, 42 (February
1996): What did we learn from the comet crash?

Morfill, *Nature,* **381**, 279 (May 23, 1996): Have we finally understood
Jupiter's rings?

Landbury, *Physics Today,* 7/1996, 17 and 19: Findings of the probe/
Io's core

Kerr, *Science,* **272**, 1589 (June 14, 1996): What is driving Jupiter's
winds?

Kerr, *Science,* **273**, 311 (July 19, 1996): Ganymede's magnetosphere

Stevenson, *Nature,* **384**, 511 (December 12, 1996): Ganymede's earlier core dynamo

McKinnon, *Nature,* **386**, 765 (April 24, 1997): Evidence for an ocean
on Europa

Kerr, *Science,* **276**, 1648 (June 13, 1997): Why is Callisto so inactive?

McKinnon, *Nature,* **390**, 23 (November 6, 1997): The four Galilean
moons all together

Kerr, *Science,* **279**, 30 (January 2, 1998): Oceans on Europa—and
Callisto?

Important Articles in Technical Journals

(available from university libraries)

GRL = Geophysical Review Letters

JGR = Journal for Geophysical Research

SSR = Space Science Reviews

Special issues devoted to the Pioneer, Voyager, and Galileo Missions:

Pioneer:

Science, **204**, 945 (1979) and **206**, 925 (1979), as well as *Nature,* **280**,
725 (1979): First events

GRL, 7, 1 (1980) and JGR, 86, 8123 (1981): Further analysis

Galileo:

Science, **253**, 1516 (1991): Galileo at Venus

SSR, 60, 3 (1992): Detailed description of the Galileo mission (more
than 600 pages!)

Icarus, **107**, 2 (1994): Galileo at Gaspra

Icarus, **120**, 1 (March 1996): Galileo at Ida

Science, **272**, 837 (May 10, 1996): The Galileo probe's initial findings

Science, **274**, 377 (October 18, 1996): The Galileo orbiter's initial
findings from Jupiter

Further important articles about individual Galileo findings

Belton et al., *Science,* **257**, 1647 (September 18, 1992): The first photographs of Gaspra

Sagan et al., *Nature,* **365**, 715 (October 21, 1993): How Galileo
searched for life on Earth

Belton et al., *Science,* **256**, 1543 (September 9, 1994): The first photographs of Ida

Chapman et al., *Nature,* **374**, 783, and Belton et al., ibid., 785 (April
27, 1995): The discovery of Ida's moon Dactyl and the determina-
tion of the density of Ida

Anderson et al., *Science,* 272, 709 (May 3, 1996): Gravimetry of Io

Pryor et al., GRL, 23, 1893 (July 15, 1996): EUVS readings of the interplanetary Lyman-alpha luminescence

Kivelson et al., *Science,* 273, 337 (July 19, 1996): A magnetic field around Io?

Grün et al., *Science,* 274, 399 (October 18, 1996): Dust readings during Galileo's approach

Gurnett et al., *Nature,* 384, 535, Kivelson et al., ibid., 537, Anderson et al., ibid., 541, and Schubert et al., ibid., 544 (December 12, 1006): The discovery of Ganymede's magnetosphere and magnetic field, as well as its gravimetry

Gurnett et al., *Nature,* 387, 261, Khurana et al., ibid., 262, and Anderson et al., ibid., 264 (May 15, 1987): Callisto: no magnetic field and no differentiation.

Anderson et al., *Science,* 277, 1236 and Kivelson et al., ibid., 1239 (May 23, 1997): Gravimetry and magnetic field readings from Europa

Kliore et al., *Science,* 277, 355 (July 18, 1997): Proof of an ionosphere on Europa

Atkinson et al., *Nature,* 388, 649, and Seiff et al., ibid., 650 (August 14, 1997): Revised probe wind speeds

Hinson et al., GRL, 24. 2107 (September 1, 1997): Jupiter's ionosphere

Kurth et al., GRL, 24, 2167 (September 1, 1997): Ganymede—a new radio source

Grün et al., GRL, 24, 2171, and Horanyi et al., ibid., 2175 (September 1, 1997): Dust readings in the Jovian magnetosphere and their significance

Khurana et al., GRL, 24, 2391 (October 1, 1997): Indications that Io has a magnetosphere of its own

Lopes-Gautier et al., GRL, 24, 2439, and McEwen et al., ibid., 2443 (October 15, 1997): NIMS and SSI results on Io's volcanism

Habbal et al., *Astrophysical Journal,* **489**, L103 (November 1, 1997): Galileo as an observer of the solar corona

Carr et al., *Nature,* **391**, 363 (January 22, 1998) and three further articles: Evidence for an ocean on Europa?

Chyba, *Nature,* **403**, 381 (January 27, 2000): Energy for microbial life on Europa

Gierasch et al., *Nature,* **403**, 628, and Ingersoll et al., ibid., 630 (February 10, 2000): Moist convection in Jupiter's atmosphere

Graps et al., *Nature,* **405**, 48 (May 4, 2000): Io as a source of the Jovian dust streams

Gaidos & Nimmo, *Nature,* **405**, 637 (June 8, 2000): Tectonics and water on Europa

Prockter & Pappalardo, *Science,* **289**, 941 (August 11, 2000): Folds on Europa: Implications for crustal cycling

Kivelson et al., *Science,* **289**, 1340 (August 25, 2000): Galileo magnetometer measurements: a stronger case for a subsurface ocean at Europa

Showman & Dowling, *Science,* **289**, 1737 (September 8, 2000): Nonlinear simulations of Jupiter's 5-micron hot spots

Index